SpringerBriefs in Applied Sciences and Technology

SpringerBriefs present concise summaries of cutting-edge research and practical applications across a wide spectrum of fields. Featuring compact volumes of 50 to 125 pages, the series covers a range of content from professional to academic.

Typical publications can be:

- A timely report of state-of-the art methods
- An introduction to or a manual for the application of mathematical or computer techniques
- A bridge between new research results, as published in journal articles
- A snapshot of a hot or emerging topic
- An in-depth case study
- A presentation of core concepts that students must understand in order to make independent contributions

SpringerBriefs are characterized by fast, global electronic dissemination, standard publishing contracts, standardized manuscript preparation and formatting guidelines, and expedited production schedules.

On the one hand, **SpringerBriefs in Applied Sciences and Technology** are devoted to the publication of fundamentals and applications within the different classical engineering disciplines as well as in interdisciplinary fields that recently emerged between these areas. On the other hand, as the boundary separating fundamental research and applied technology is more and more dissolving, this series is particularly open to trans-disciplinary topics between fundamental science and engineering.

Indexed by EI-Compendex, SCOPUS and Springerlink.

Muhamad Husaini Abu Bakar ·
Tajul Adli Abdul Razak · Andreas Öchsner
Editors

IT Solutions for Sustainable Living

 Springer

Editors
Muhamad Husaini Abu Bakar
Malaysian Spanish Institute
Universiti Kuala Lumpur
Kulim, Kedah, Malaysia

Tajul Adli Abdul Razak
Malaysian Spanish Institute
Universiti Kuala Lumpur
Kulim, Kedah, Malaysia

Andreas Öchsner
Faculty of Mechanical Engineering
Esslingen University of Applied Sciences
Esslingen am Neckar, Baden-Württemberg,
Germany

ISSN 2191-530X   ISSN 2191-5318 (electronic)
SpringerBriefs in Applied Sciences and Technology
ISBN 978-3-031-51858-4   ISBN 978-3-031-51859-1 (eBook)
https://doi.org/10.1007/978-3-031-51859-1

This Springer imprint is published by the registered company Springer Nature Switzerland AG
The registered company address is: Gewerbestrasse 11, 6330 Cham, Switzerland

Paper in this product is recyclable.

Preface

In today's dynamic and competitive business environment, industries face numerous challenges related to process optimization, energy consumption, and technological advancements. Efficient management of industrial processes is crucial for maximizing productivity, reducing waste, and ensuring sustainable growth. Similarly, energy consumption and maintenance play a pivotal role in achieving cost-effectiveness and environmental sustainability. Moreover, advancements in technology and engineering have the potential to revolutionize industrial operations and enhance overall performance. This research book provides an analysis of various topics encompassing industrial processes, energy consumption, and technology in engineering management. The book encompasses three main categories. The first category is related to industrial processes, which investigate methodologies and techniques to identify and eliminate waste. The second category discusses energy consumption and maintenance strategies in industrial and building environments. This section aims to optimize energy usage, reduce costs, and promote sustainable practices. Lastly, the third category in technology and engineering explores advancements in technology and engineering and their applications in industrial settings. This section investigates innovative solutions and techniques to enhance productivity, optimize processes, and improve overall performance. The research book aims to become a valuable resource for researchers, practitioners, and policymakers interested in smart technologies and sustainable living.

Kulim, Malaysia   Muhamad Husaini Abu Bakar
Kulim, Malaysia   Tajul Adli Abdul Razak
Esslingen am Neckar, Germany   Andreas Öchsner

Contents

Wastes Detection Using a Value Stream Mapping (VSM) Technique in an Electronic-Based Manufacturing Company

Daniel Ikmal Abu Hassan, Mohd Norzaimi Che Ani, Shahrul Kamaruddin, and Ishak Abdul Azid

Abstract Production wastes consist of inventory, motion, transportation, waiting, over-processing, over-production, and waiting. Improper management of these identified wastes causes interruption of the production system due to unplanned stoppage time, quality issues, material delinquencies, and other related issues. Thus, in this research, wastes hunting activities were conducted to identify the production wastes using a proper method known as the value stream mapping (VSM). This research was conducted focusing on an electronic-based company because their finished good and product required hundreds of small components and most of the potential wastes were hidden due to the fact that the assembly process is using either a semi-automation or fully automation process. The objective of this research was to define the proper way of the waste detection process to create the opportunity for continuous improvement activity. The result obtained showed that the percentage of wastes was 95%, and this situation appeared as hidden waste causing the issues in the production system.

Keywords Production wastes · Production interruption · Wastes hunting · Value stream mapping (VSM) · Electronic-based

D. I. Abu Hassan · M. N. Che Ani (✉)
Manufacturing Section, Universiti Kuala Lumpur, Malaysian Spanish Institute, Kulim Hi-Tech Park, 09000 Kulim, Kedah, Malaysia
e-mail: mnorzaimi@unikl.edu.my

S. Kamaruddin
Engineering Section, Universiti Kuala Lumpur, Malaysian Spanish Institute, Kulim Hi-Tech Park, 09000 Kulim, Kedah, Malaysia
e-mail: shahrul.k@utp.edu.my

I. A. Azid
Mechanical Engineering Department, Universiti Teknologi Petronas, 32610 Bandar Seri Iskandar, Perak, Malaysia
e-mail: ishak.abdulazid@unikl.edu.my

1 Introduction

Production performance, efficiency, and productivity commonly are interrupted by several issues such as product defects, production stoppage time due to machine downtime or absence of the workers, incoming material issue, and many more issues related to either machines, materials, methods of process, or workers. All these interruption elements appeared due to poor management, hidden or unrealized wastes of the production process. Ohno [7] defined in the management of the production system, the seven types of wastes must be clearly identified to ensure the production performance achieving the maximum rate. The seven types of wastes have been recognized by the Toyota production system (TPS) through the lean manufacturing (LM), known as inventory, motion, transportation, waiting, over-processing, over-production, and waiting [6]. Improper management of these identified wastes caused unplanned stoppage time, quality issues, material delinquencies, and other related issues.

Since LM is widely recognized by the worldwide manufacturing industries for improving the production performance, efficiency, and productivity through eliminating or minimizing wastes but still room of improvements are available [4]. In LM, the first step is the elimination of wastes in production floor then maximizing the production flow by using appropriate lean tools to ensure the production become more effective and efficient [9]. Currently, LM is widely applied as a continuous improvement technique in the long-term planning for the purpose meeting customer requirements through wastes elimination [3]. In current scenario, many organizations in the worldwide of manufacturing industries are adopting the continuous improvement culture to improve their production system [1]. The implementation of LM is aimed for the organizations to remain competitive in the local and global markets [2]. The continuous improvement culture focuses on the eliminating or minimizing the production interruption issues. Since the seven types of wastes have been recognized by the LM system, the value stream mapping (VSM) has been identified as a powerful tool of the wastes detection.

Practically, the implementation of continuous improvement activity normally starts with screening the condition of production activities using VSM, and then all the possible wastes will be identified from mapping before creating the corrective action to eliminate or minimize the wastes [5]. Most of the wastes detection activity using VSM was successfully implemented in the automotive-based industry through the LM system [8], but the implementation in other industries such as electronics-based might differ due to the working culture, production facilities, and the environment. Hence, this research was conducted by focusing on the electronic-based company because their finished good and product required hundreds of small components, and most of the potential wastes were hidden due to the fact that assembly process is using either a semi-automation or fully automation process. The objective of this research was to define the proper way of the wastes detection process to create the opportunity for continuous improvement activity.

The focus of this research is to discuss the implementation of wastes detection using VSM in an electronics-based industry. The first objective of this research is to identify the suitability of suitable steps of implementation of VSM in the electronics-based industry. The percentage of the production wastes obtained from VSM will be evaluated and interpreted as the second objective which is to indicate the level of the production performance against the non-value-added activities. The final objective is to discuss the possible corrective action that must be implemented as an obtained result from the implementation of VSM to improve toward smoothness and efficient production system. Section 2 of this research describes the methodology of the VSM implementation in the electronics-based industry, then followed by Sect. 3 which discusses the implementation of the VSM result. Later, the findings in VSM implementation are discussed in Sect. 4. Finally, the conclusion of this research will be provided in Sect. 5.

2 The Methodology of the Value Stream Mapping Technique

Commonly, the screening of the production wastes in the production system is widely employed through the VSM technique. As mentioned in the introduction, VSM was originally initiated in the LM as implemented by TPS, and it was quickly spread into other automotive manufacturing industries because of the effectiveness of the wastes detection activity [10]. Currently, most of the industries either electronic-based or others also adopted the implementation of the VSM to identify the wastes in their production system. Implementation of the VSM can be divided into four stages which are product identification, component identification, Gemba-walk, and VSM current state as summarized in Fig. 1.

In the electronics-based industry, hundreds of product variances are available in their listed products. The fluctuation demands of each variance normally happen, especially in semiconductor or contract manufacturer electronics companies. This is because their production scheduling totally depends on the customer's request. Based on this situation, the implementation of the VSM will start with a product selection

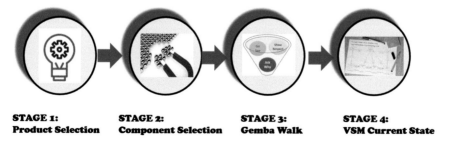

STAGE 1: **Product Selection**   **STAGE 2:** **Component Selection**   **STAGE 3:** **Gemba Walk**   **STAGE 4:** **VSM Current State**

Fig. 1 Methodology of the value stream mapping technique

in stage 1. A product will be selected based on the highest demand because it will hugely impact the overall performance of the production system. Once the product has been selected, stage 2 consists of the selection of the components. Each product required multiple components to be assembled and transformed into a finished good and product. In VSM, the component will be selected based on the involvement of the component of the entire production process.

A Gemba-walk will be conducted in stage 3 which means that all the related process flows will be observed in terms of process cycle time and inventory of each process. All the observed data will be recorded based on the observed time to ensure the accuracy of the data collection and observation. Upon completion of stage 3, then the VSM current state will be drawn based on recorded data during the Gemba-walk activity. All the related standard symbols will be used in drawing of VSM's current state. In addition, the percentage of the value-added ratio (var) will be calculated by using Eq. (1) to represent the situation of the wastes in the observed production system.

$$\text{Value - added ratio(var)} = [\text{value - added time/lead time}]\% \tag{1}$$

Finally, once the production wastes status has been defined in the VSM current state, the potential corrective action will be suggested to improve the production performance, availability, and quality. This VSM will help the management of the organization to detect causes of the production's low performance, unplanned stoppage time, or process idling.

3   Data Collection

This section presents and discusses the detailed process of production wastes detection using the VSM technique as applied in selected contract manufacturer of the electronic-based industry. The case study company is involved in contract manufacturing of printed circuit board (PCB) assemblies. The current production layout is set up according to the product-based production system which is a series links production chain layout. The straight-line production layout consists of four assembly processes, and each process is managed by a different dedicated worker. Since the product-based production layout is employed in this case study, the performance of the production system will be measured based on the daily production output, and based on the initial observation, there is a lot of wear and tear interruption during operation of the production process.

The VSM technique had been implemented in this selected case study company to identify the potential wastes interrupting the production process. The step-by-step of the VSM implementation as discussed in the previous section had been carefully followed. The obtained result of the VSM current state had been drawn based on the data analysis and evaluation as shown in Fig. 2.

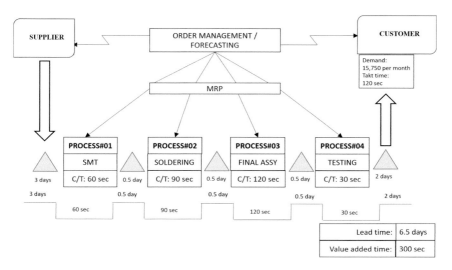

Fig. 2 VSM current state

Product XX had been selected from the case study industry, and this product required four processes to complete the finished-good product. The monthly demand of the product XX was 15,750 units per month, and it was converted into a takt time which was 120 s. The conversion into takt time was based on 21 working hours per day and 25 working days per month. During the Gemba-walk, inventory level had been counted and converted into the stock per day, and it was found that the total lead time was 6.5 days, while the value-added time or processing time only required 300 s per unit.

Based on the obtained results of the production lead time and value-added time, then the value-added ratio (var) had been calculated using Eq. (1). The obtained result of var was 0.053%. That means the percentage of the production wastes in this selected product was 99.947%. The production wastes occurred because of the high level and unstandardized inventory between the processes and unbalanced process cycle time. In the VSM future state, recommendations were made to improve these identified wastes toward the optimum performance of the production system.

4 Finding and Discussion

Value-stream mapping or sometimes called "material and information flow mapping" is a method of screening the production activities and wastes detection by analyzing the current activities throughout the entire related processes on the production floor. The activities of observation will be evaluated and analyzed starting from receiving raw materials from suppliers until accepted by customer. The outcomes of the VSM implementation are to visualize the entire production flow and display all activities using a specific diagram. The core value of the VSM technique is that both materials and information from entire process will be captured and recorded. From the VSM's current state, all the wastes will be identified, and the potential corrective action will be suggested for the continuous improvement activities.

In the wastes detection activity on the production floor, the required first step is to translate the current observed activity into VSM current state diagram. This diagram can help the organization identify the potential wastes that interrupt the production flow such as process idling, unbalanced process time, unrequired inventory, and other potential wastes. These identified wastes will then be minimized or eliminated during the corrective action plan to help streamline the production process and indirectly achieving the optimum production efficiency. VSM is the technique of using diagrams with dedicated symbols to illustrate the flow of a process and to identify opportunities for improvement. The VSM technique allows organizations to create a detailed visualization of all required steps as practices on the production floor by evaluating all the information especially process cycle time and inventory level. The diagram of the VSM represents the production flow process from supplier to the customer throughout the entire production process and to identify the areas of improvement opportunities.

5 Conclusions

In this paper, the application of the VSM technique is an electronics-based company or specifically the contract manufacturer of the electronic assembly process has been successfully analyzed and evaluated. Based on the results obtained from VSM implementation, there were two main significant production wastes that interrupted the production performance which were identified as excess inventory and unbalanced process cycle time. The obtained result from the implementation of the VSM found the value-added ratio (var) of 0.053% or the percentage of the production wastes was 99.947%. The VSM was successfully implemented, and the area of improvement opportunities was identified based on the visualized VSM diagram. It is highly recommended in the VSM future state to address these identified wastes toward the optimum performance of the production system and indirectly minimize the production lead time. Based on the implementation, it is foreseen that prior to implementing the continuous improvement activities, the screening of the wastes or any production issues must be implemented and VSM is the most suitable and reliable technique for the production wastes hunting activities.

Acknowledgements This paper is fully supported by finance from Universiti Kuala Lumpur (UniKL) and STRG grant (UniKL/CoRI/str20014). Appreciation is also extended to the selected case study industry and anonymous reviewers for the comments given which led to the significantly improved manuscript quality.

References

1. M.N.C. Ani, S. Kamaruddin, I.A. Azid, Leanness production system through improving of upstream process based on check-act-plan-do cycle. Int. J. Six Sigma Compet. Advantage **11**(2–3), 95–113 (2019)
2. F. Behrouzi, K.Y. Wong, Lean performance evaluation of manufacturing systems: a dynamic and innovative approach. Procedia Comput. Sci. **3**(1), 388–395 (2011)
3. S. Bhasin, P. Burcher, Lean viewed as a philosophy. J. Manuf. Technol. Manag. **17**(1), 56–72 (2006)
4. M.N. Che Ani, J.F. Chin, Self-reinforcing mechanisms for cellularisation: a longitudinal case study. Int. J. Prod. Res. **54**(3), 696–711 (2016)
5. M. Emiliani, D. Stec, Using value-stream maps to improve leadership. Leadersh. Organ. Dev. J. **25**(8), 622–645 (2004)
6. J.K. Liker, J.M. Morgan, The Toyota way in services: the case of lean product development. Acad. Manag. Perspect. **20**(2), 5–20 (2006)
7. T. Ohno, (2011) How the Toyota Production System was created. in The Anatomy of Japanese Business, vol 10(4) (Routledge, New York), pp. 133–144
8. M.P. Pérez, A.M. Sánchez, Lean production and supplier relations: a survey of practices in the Aragonese automotive industry. Technovation **20**(12), 665–676 (2000)
9. M. Scherrer-Rathje, T.A. Boyle, P. Deflorin, Lean, take two! Reflections from the second attempt at lean implementation. Bus. Horiz. **52**(1), 79–88 (2009)
10. U.K. Teichgräber, M. de Bucourt, Applying value stream mapping techniques to eliminate non-value-added waste for the procurement of endovascular stents. Eur. J. Radiol. **81**(1), e47–e52 (2012)

A Review of Six Sigma Implementation in Manufacturing Industries

Nordini Hannani Mohd Asri, Mohd Norzaimi Che Ani, Shahrul Kamaruddin, and Ishak Abdul Azid

Abstract This paper presents a review of the Six Sigma concept and applications employed in manufacturing industries that were analyzed based on published articles. The Six Sigma concept is widely recognized and utilized by worldwide organizations, especially in continuous improvement activities solving the production issues. This research was conducted by chosen articles related to the implementation of the Six Sigma in selected case study industries. The main objectives of this research are to identify the success factors of Six Sigma implementation, to define the suitable tools applied in the Six Sigma methodology, and to identify the constraints and limitations of the Six Sigma implementation. The outcomes obtained from the review articles showed that even though the implementation of Six Sigma is generally similar in terms of methodology, the application of the tools is significantly different based on the problem statement, and the completion time is also different due to complexity of the selected process.

Keywords Six Sigma · Manufacturing industries · Continuous improvement · Production issues · Tools

N. H. Mohd Asri · M. N. Che Ani (✉)
Manufacturing Section, Universiti Kuala Lumpur, Malaysian Spanish Institute, Kulim Hi-Tech Park, 09000 Kulim, Kedah, Malaysia
e-mail: mnorzaimi@unikl.edu.my

N. H. Mohd Asri
e-mail: nordini.asri@s.unikl.edu.my

S. Kamaruddin
Mechanical Engineering Department, Universiti Teknologi Petronas, 32610 Bandar Seri Iskandar, Perak, Malaysia
e-mail: shahrul.k@utp.edu.my

I. A. Azid
Engineering Section, Universiti Kuala Lumpur, Malaysian Spanish Institute, Kulim Hi-Tech Park, 09000 Kulim, Kedah, Malaysia
e-mail: ishak.abdulazid@unikl.edu.my

© The Author(s), under exclusive license to Springer Nature Switzerland AG 2024
M. H. Abu Bakar et al. (eds.), *IT Solutions for Sustainable Living*,
SpringerBriefs in Applied Sciences and Technology,
https://doi.org/10.1007/978-3-031-51859-1_2

1 Introduction

Quality is the main element in satisfying customers and retaining their loyalty to continue business engagement with any organization. The meaning quality is fulfilling the customer needs and requirements, and products that have a significant impact on quality will remain competitive in the global markets for long-term revenue and profitability [1]. That is the reason the continuous improvement culture is critical for any organization, especially for quality enhancement or improvement. One of the most critical quality issues is a variation of the processed product and causing a poor product quality due to out-of-specification dimensions. Based on this situation, Motorola, one of the biggest industrial players, has successfully initiated and developed the systematic quality improvement plan, known as the Six Sigma model with the application of define-measure-analyze-improve-control (DMAIC) methodology in response to poor quality and customer complaints, which harmed the company's competitiveness [3]. The Six Sigma model consists of integration between the statistical analysis and application of the quality tools to overcome the issues of the poor product quality in the production system [10]. Currently, the DMAIC methodology in the Six Sigma model is recognized by many industrial players, and it has been used in many organizations worldwide.

Six Sigma has been named as a quality improvement strategy because the focused element is processed variation of six standard deviations (6σ) of acceptable limits between the specification limits consisting of lower specification limit (LSL) and upper specification limit (USL) [7]. Specification limits are normally derived from the customer requirements, and the specification tolerances will be determined by the LSL and USL. Then, the internal monitoring of the process variation between two control limits known as lower control limit (LCL) and upper control limit (UCL) will be evaluated. In Six Sigma, the thinness of the control limits will provide the good impacts of the process variation toward achieving good quality of the product [9]. To solve a problem of process variation, the Six Sigma model will identify the root cause and then eliminate it through five whys basic root cause analysis technique known as DMAIC methodology [8]. Six Sigma has been widely implemented with the main objectives which are to improve the finished good and product quality by identifying and solving the main root causes of defects or errors in the production floor.

Since the Six Sigma is one of the successful models for improving business through quality improvement activity, many organizations have adopted the Six Sigma model in their organizations with the main objective of improving product quality [4]. Six Sigma is not a dedicated model of implementation in selected industries only but it's flexible and robust due to the application of Six Sigma in various organizations is different due to the different environments, cultures, ethics, and nature of business. Hence, this research attempts to review the present status of Six Sigma and deliberate the different techniques of applying Sigma in different organizations of manufacturing industries.

The main focus of this research is to discuss the implementation of Six Sigma in manufacturing industries based on obtained results from various literatures. The primary objectives of this research are to identify the success factors of Six Sigma implementation, to define the suitable tools that will be applied in the Six Sigma methodology, and to identify the constraints and limitations of the Six Sigma implementation. The following section describes the research approach, followed by results and analysis of the implementations of Six Sigma methodology. Then, the findings will be discussed in the following section, and the conclusion will be concluded at the end of this paper.

2 Inductive Research Approach

In order to study the successful implementation and the constraints and limitations of the Six Sigma implementation in manufacturing industries, a broad inductive research approach is adopted. Firstly, it is to systematically synthesize the previous research works that have been conducted based on the Six Sigma implementation from the published literatures. Secondly, a comprehensible identification of the tools and techniques throughout the DMAIC methodology in the Six Sigma model has been established. Finally, the constraints and limitations of Six Sigma have been evaluated based on the outcomes from previous works.

The methodology of this research as shown in Fig. 1 is carried out by exploring the literature related to the theory and implementations of Six Sigma as applied in manufacturing industries that are available in various publications such as journals and conference proceedings. The selection of the related articles is based on the assessment of the Six Sigma implementation themes. This study is limited to articles related to Six Sigma implementation in manufacturing industries only because of different nature of the establishment and working environment of manufacturing industries using a similar Six Sigma model to improve their product quality.

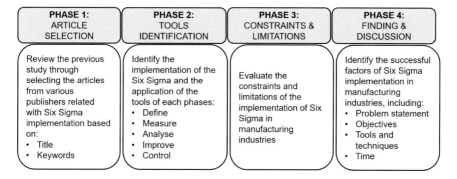

Fig. 1 Inductive research approach

As the methodology shown in Fig. 1, the previous works of Six Sigma implementation were selected based on the title and keywords. Once articles were selected, the application of the continuous improvement tools for each phase in Six Sigma methodology was identified. Then, the constraints and limitations were evaluated based on a case study of the selected articles. Finally, the obtained results from the review of the previous works were discussed based on the identified problem statement, defined objectives, the obtained results, and the duration of the Six Sigma implementation.

3   The Results of Data Collection

As mentioned in previous sections, Six Sigma is now being fully stretched, and it has been applied in many organizations of manufacturing industries. From published journals, there are many well-documented examples of the Six Sigma implementation in manufacturing industries. Even though Six Sigma was originally introduced in the electronics industry, nonetheless in the present time, the majority of industries have been implementing Six Sigma to gain the maximum benefits of its implementation and create a continuous improvement culture. This section presents and discusses the detailed approach of Six Sigma as applied by the manufacturing industries.

The review and analysis of the selected articles showed the suitability of Six Sigma implementation in manufacturing industries as an improvement approach for solving the issue of the poor quality to meet customer expectation. Currently, Six Sigma is widely accepted in all types of organizations in manufacturing industries. There are significant differences in terms of Six Sigma tools and approaches used in the DMAIC methodology. Even though the application of Six Sigma has similar objectives, the approaches are slightly different based on the identified problem statement and defined objective. The obtained results of the analysis and evaluation of the previous works are tabulated in Table 1.

4   Finding and Discussion

The findings result from selected articles showed the suitability of Six Sigma as an improvement approach for solving the quality issues in production system. Practically, the Six Sigma implementation normally starts with defining the problem statement, and then all the possible problems will be measured based on the current situation in production floor. Before any improvement activities are carried out, the organization will analyze the identified problems to ensure proper solutions to the problem in the improve phase. Any problems will be solved systematically by using proper quality tools and techniques to eliminate the root cause in order to achieve the defined objective. Once successfully implemented the improvement process,

Table 1 Results of the data collection

Author(s)	Research background	Application of tools and techniques				
		Define	Measure	Analyze	Improve	Control
Kaushik et al. [6]	High rejection rate of precision process, and the objective is reducing defects inherent in the processes	Process map SIPOC	Gauge R&R	Process capability Fishbone diagram	Two-sample t-test	Xbar-R chart
Wang and Chen [11]	Improvement in the supportive activities of an equipment manufacturer	Process map SIPOC	I-chart	Pareto diagram FMEA	Relation diagram	Xbar-R chart
Zhang et al. [12]	Improve the process capability to meet the thickness requirements	Yield analysis	Gauge R&R Process map	Fishbone diagram C&E matrix	FMEA ANOVA Scatter plot Regression analysis	IMR-chart
Gijo et al. [5]	Reduction of defects in the machining process	Critical to Quality (CTQ) Process map SIPOC	Gauge R&R	Pareto chart Fishbone diagram Process capability	DOE	Cause–Solution matrix Xbar-R chart
Ani et al. [2]	Reduction of defects of the manual process	5W1H SIPOC Scatter plot	Pareto chart Process map Process capability	Fishbone diagram Multi-voting analysis	Hypothesis test	Scatter plot Process capability

then that activity will be controlled to ensure the sustainability of the improvement program.

In terms of the application of Six Sigma, the application of the continuous improvement tools does not focus on dedicated tools for each DMAIC phase but multiple numbers of Six Sigma tools that have been applied. The applicability of the tools such as process map, fishbone diagram, Pareto chart, process capability, ANOVA, DOE, control chart, and many more is dependent on the working environment, culture, ethics, and nature of the production system. The applications of Six Sigma tools are aimed to achieve the objectives of each phase and primarily achieve the main objective which is improving the quality issue. From the results of the reviewed articles, the implementation of Six Sigma generally requires more than six months for the completion of the DMAIC phases. It is considered as the model for the medium- to long-term corrective action of solving the quality issue. The implementation of Six Sigma with the selected quality tools and techniques will provide a high impact of continuous improvement culture to the organization.

5 Conclusions

In this paper, the implementation of the Six Sigma technique focusing in the production floor of manufacturing industries has been analyzed through a review of published articles. Based on the results obtained from reviewing the articles, Six Sigma has been recognized and implemented by various manufacturing industries to solve the issue of the poor product quality of their production system. Even though the implementation of Six Sigma has a similar objective, the application of the continuous improvement tools in every phase of the DMAIC methodology is slightly different. The flexibility of the selected tools and techniques is depending on the identified problem statement and the defined objective of the case study. The tools and techniques can be assigned to any phase of DMAIC and can be repeatedly used in different phases. There are no limits to the quality tools and techniques, and any tools will be applied according to the suitability of the situation. In terms of implementation duration, most of the organizations took more than six months to establish the Six Sigma model to solve the quality issue in their production system. Overall, the Six Sigma implementation, especially in manufacturing industries basically applied similar objectives and, in some cases, slightly different tools to suit the problems encountered and the different environments.

Acknowledgements This paper is fully supported by finance from Universiti Kuala Lumpur (UniKL) and STRG grant (UniKL/CoRI/str20014). Appreciation is also extended to the selected case study industry and anonymous reviewers for the comments given which led to the significantly improved manuscript quality.

References

1. B. Angelova, J. Zekiri, Measuring customer satisfaction with service quality using American customer satisfaction model (ACSI Model). Int. J. Acad. Res. Bus. Soc. Sci. **1**(3), 232 (2011)
2. M.N.C. Ani, I.A. Azid, S. Kamaruddin, Solving quality issues in automotive component manufacturing environment by utilizing six sigma DMAIC approach and quality tools. Int. Conf. Ind. Eng. Oper. Manage. **1**(1), 8–10 (2016)
3. M. Barney, Motorola's second generation. Six Sigma Forum Mag. **1**(3), 13–16 (2002)
4. E. Drohomeretski, S.E. Gouvea da Costa, E. Pinheiro de Lima, P.A.D.R. Garbuio, Lean, Six Sigma and Lean Six Sigma: an analysis based on operations strategy. Int. J. Prod. Res. **52**(3), 804–824 (2014)

5. E.V. Gijo, J. Scaria, J. Antony, Application of Six Sigma methodology to reduce defects of a grinding process. Qual. Reliab. Eng. Int. **27**(8), 1221–1234 (2011)
6. P. Kaushik, D. Khanduja, K. Mittal, P. Jaglan, A case study: application of Six Sigma methodology in a small and medium-sized manufacturing enterprise. Total Qual. Manage. **24**(1), 4–16 (2012)
7. P.N. Koch, R.J. Yang, L. Gu, Design for six sigma through robust optimization. Struct. Multidiscip. Optim. **26**(3), 235–248 (2004)
8. M.C. Lee, T. Chang, Combination of theory of constraints, root cause analysis and Six Sigma for quality improvement framework. Int. J. Prod. Qual. Manage. **10**(4), 447–463 (2012)
9. B.M.M. Prabhuswamy, Process variability reduction through statistical process control for quality improvement. Int. J. Qual. Res. **642**(1), 727–740 (2011)
10. M.S. Raisinghani, H. Ette, R. Pierce, G. Cannon, P. Daripaly, Six Sigma: concepts, tools, and applications. Ind. Manag. Data Syst. **105**(4), 491–505 (2005)
11. F.K. Wang, K.S. Chen, Application of Lean Six Sigma to a panel equipment manufacturer. Total Qual. Manag. Bus. Excell.Excell. **23**(3–4), 417–429 (2012)
12. M. Zhang, W. Wang, T.N. Goh, Z. He, Comprehensive Six Sigma application: a case study. Prod. Plan Control **26**(3), 219–234 (2015)

Risk Assessment for Solving the Production Variation in Small Medium-Sized Enterprises

Mohd Norzaimi Che Ani and Ishak Abdul Azid

Abstract This paper presents a risk assessment for solving the production variation in small medium-sized enterprises (SME). In industrial practice, the risk issue, unsafe working condition, and workplace accidents have increased concurrently with the industrial development. The risk assessment becomes a mandatory process in producing the finished good product in the production system. Failure mode and effects analysis (FMEA) is widely used to identify and solve the potential risk in the production floor. In the FMEA, the risk priority number (RPN) is calculated through the multiplication of rating given to potential risk occurrence of different hierarchical levels of failure modes. Thus, in this paper, the main objective is to identify and investigate the selected processes in SME. This risk assessment will detect the potential failure of the production system and indirectly can be predicted upfront. Therefore, the production risk assessment can be performed and improve the effectiveness and efficiency of the production flow of the SME.

Keywords Small medium-sized enterprise (SME) · Failure mode and effects analysis (FMEA) · Risk priority number (RPN) · Risk assessment

1 Introduction

In industrial practice, the risk issue, unsafe working condition, and workplace accidents have increased concurrently with industrial development, especially towards Industry 4.0. At the same time, the risk assessment becomes a mandatory process in

M. N. Che Ani (✉)
Manufacturing Section, Universiti Kuala Lumpur Malaysian Spanish Institute, Kulim Hi-Tech Park, 09000 Kulim, Kedah, Malaysia
e-mail: mnorzaimi@unikl.edu.my

I. A. Azid
Mechanical Section, Universiti Kuala Lumpur Malaysian Spanish Institute, Kulim Hi-Tech Park, 09000 Kulim, Kedah, Malaysia
e-mail: ishak.abdulazid@unikl.edu.my

© The Author(s), under exclusive license to Springer Nature Switzerland AG 2024
M. H. Abu Bakar et al. (eds.), *IT Solutions for Sustainable Living*,
SpringerBriefs in Applied Sciences and Technology,
https://doi.org/10.1007/978-3-031-51859-1_3

producing the finished good product in the production system. Currently, the quality management system (QMS) has been revised for the policy of ISO9001 standard which is its current version (ISO9001:2015) has emphasized that risk assessment is one of the key elements in developing the QMS for any organizations as stated in Clause 6.1: Actions to address risks and opportunities. The same thing applies to industries in the automotive sector, where the International Automotive Task Force 16949:2016 (IATF 16949:2016) has enforced the compulsory of six core tools in the automotive assembly process towards meeting the automotive compliance. One of the core tools is the failure mode and effects analysis (FMEA) to identify and solve the potential risk in the design stage or production stage. Based on the review for both standards (ISO9001:2015 and IATF16949:2016), the risk assessment is strongly related to the development of Industry 4.0 or known as digital manufacturing.

Digital manufacturing adopts the Internet of things (IoT) between process to process (P2P) in the production floor towards effective and efficient production system and failure of designing a smooth production system due to occurring risks that will impact the development of Industry 4.0. In the FMEA analysis, the risk priority number (RPN) is calculated through the multiplication of the rating given to the potential risk occurrence of different hierarchical levels of failure modes. The identification of severity (S), occurrence (O), and detection (D) is based on ratings 1 to 10, where rating 1 is a low risk and 10 is a high risk. At present, the risk assessment in actual practice, especially in the SMEs, is not effective as it has not been able to quantify the potential production risk due to fact that the identification and measurement of the risk is based on experiences and mutual consent among the employees involved in the production process with many assumptions made in the rating index. Based on current practices, the risk identification of each element in the RPN calculation is questionable because the determination index is unable to quantify the actual potential risk and sometimes does not reflect the actual failure mode risk of practices.

Thus, in this paper, the main objective is to identify and investigate the selected processes in SME. This risk assessment will detect the potential failure of the production system and indirectly can be predicted upfront to ensure smoothness of P2P connection in production systems. The individual elements of the production risk assessment will be identified and investigated for each process in the production system based on the sequence of P2P. The P2P of the production system and the critical factor of the production risk assessment element will be quantified based on a ten-point numeric. Therefore, the production risk assessment can be performed, and the P2P linkage and digitalization of the production system can establish the effectiveness and efficiency of the production flow, especially in SME.

2 Literature Review

The transition of the industrial revolution known as digital manufacturing or Industry 4.0 requires a cyber-physical system consisting of an integrated automation production system with electronics and an information technology (IT) device through the Internet of things (IoT). The smoothness of the production process is the primary objective of the Industry 4.0 revolution to ensure that the production system is producing the finished good part with optimum efficiency and fastest lead time. Adaptation to the IT system for connecting process to process (P2P) in the production processes does not only just integrate the systems but also requires stability and reliability of the baseline production system since the industrial revolution requires networked manufacturing and the concept of smart supply chains that enables the sending of product data over the IoT [7]. In terms of baseline of the model stability, several models of continuous improvement in the production system such as 4 M (man, method, material, and machine) have been introduced to maintain the baseline stability of the production system.

In the development of Industry 4.0, substantial involvement of the cyber-system requires optimum efficiency of the production system and any interruption of P2P in the production system such as waiting, unplanned downtime, machine failure, and product defect will ultimately cause failure in the development of Industry 4.0. As the industrial organization can be divided into two main categories which are large enterprise and small medium-sized enterprise (SME), the SME appears to be more struggling to meet the transition requirement of Industry 4.0 than the large enterprise. This is due to the fact that SMEs do not have the technical know-how and the capacity to successfully implement the optimum efficiency of the production system [2]. The development the Industry 4.0 for SMEs, especially in the related automotive industry is very crucial as this industry deals with thousands number of child parts. Another issue related with SME towards the development of Industry 4.0 is the readiness of industrial transition towards digital manufacturing, which is still in the infancy and questionable. This is due to the fact that most SMEs have difficulty to adopt the new transition due to several reasons such as lack of resources, finances, and comfortable with traditional system [1].

Since the development of Industry 4.0 requires stability and optimum efficiency of the production system, the risk assessment prior to developing the production system must be evaluated. The risk identification and assessment are a mandatory process and becoming current requirement for operational management [6] as stated in the standard of ISO900:2015 and IATF16949:2016. As widely practised, most of the risk assessment is carried out through employing the failure mode and effects analysis (FMEA) to evaluate the potential risk of product design and production processes [4, 5, 8]. This FMEA uses the risk priority number (RPN) based on multiplying the rating of severity (S), occurrence (O), and detection (D) and each element is identified through the estimation of severity and occurrence values [9]. However, FMEA implementation still faces some limitations, and thus, several improvement methods have been suggested in the literature to enhance the effectiveness of FMEA [3]. Pisut and Thuangporn [4] have attempted to improve a FMEA through involvement of the customer, but the analysis of FMEA was still derived based on estimation rating and mutual agreement between customer and vendor or supplier without justification rating based on quantitative manner. Liu et al. [5] developed the quantified rating by using a ten-point numeric, but it was only suitable for the independence process of production system and not for the complex processes such as process linkage in the production system. Some of the current FMEA methods are unable to handle the complex issues in terms of index rating (1–10) due to the risk rating being different based on different situations [10]. This happens because rating of S, O, and D is subjective and therefore is unable to give accurate rating.

Failure to develop the effective production system for SME will cause the critical gap between large enterprises and SME and ultimately will interrupt the development of Industry 4.0. This arising issue of alignment between both enterprises becomes big challenge for the development of Industry 4.0 since it requires synchronization of the concept of the supply chain as some of SMEs are the suppliers for large enterprise. SMEs manufacturers must take steps to mitigate these gaps through establishment of production stability by evaluating the risks of the production system with highly efficient and accurate risk assessment approach. From the production point of view, the establishment of the baseline for the development of Industry 4.0 through an effective risk assessment model does not only create opportunities to optimize the smoothness of production system but it also helps the SME to gain a competitive advantage. Thus, the risk assessment for production processes must be defined and formulated quantitatively, especially in the SME to increase the product and to enhance the readiness of the SME towards Industry 4.0.

3 Methodology

This research was carried out in three phases as shown in Fig. 1. The phase 1 consists of 2 stages where the related published literature was reviewed and analysed to understand the current trend of risk assessment for production systems. Then, the current practices of risk assessment by industrial practitioners were evaluated in selected case study industries. The results of published literature and industrial practice were synthesized to formulate the mathematical equation in phase 2. In phase 2, the mathematical formulation of risk assessment for process failure was carried out through 4 stages. The first stage identifies the risk assessment elements such as the seriousness of the failure, frequency of the failure, and other related elements. Once these elements are identified, then the risk rating was determined quantitatively through the complex relationships of process-to-process analysis. The next stage is the derivation of the equation for risk assessment through mathematical and statistical analysis. Phase 2 ended with the evaluation of the developed risk assessment formulation towards the production system based on the production output, reject rate and production efficiency. Upon completion of phase 2, the process of validation and verification was carried out in phase 3 to ensure successful formulation of risk

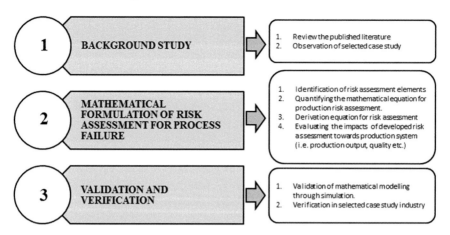

Fig. 1 Research methodology

assessment in the production system. In this phase, the validation process was carried out through the simulation process using computer software, while the verification process was carried out by the implementation of the formulated risk assessment in selected case study industries to ensure effectiveness of this research.

4 Findings and Discussion

FMEA method offered a proactive and specific method for evaluating a potential risks and failure in selected process, to identify the parts of the process that are most in need of changes due to the risks. To perform FMEA, the organization must review the steps in the process, failure modes, failure causes, and failure effects. With all these above, the organization can evaluate processes for possible failures and upfront to prevent the failure trough proactive corrective action.

FMEA is one of the key methods for every industry to determine the potential risk on specific industry so do in manufacturing industry. One of the severe risks that could happen in manufacturing industry. Quality issues after-sales causing product recall from after-market that could possibly happen if internal issues are not solve. The organization should take concern on the process which undergoes more product testing with different quality standards in the organization. This not only helps the company; it helps as well the consumer to be confident with the products. Failure to control these issues may lead to negative impact the continuation of the business by receiving less orders. Hence, drop of sales volume may affect the future direction of the organization. This emphasis on prevention may reduce risk of harm both, i.e. products and staff. FMEA is particularly useful in evaluating a new process prior to implementation and in assessing the impact of a proposed change to an existing process.

As discussed earlier in SME, inappropriate triage are processes leading to mistreatment or untimely treatment results in adverse outcome. Moreover, introduction of FMEA also helps the SME which is one of the most important industries among all in developing countries. One of the greatest risks faced by the SMEs nowadays is material shortages due to several factors. This is commonly caused by changing workforce and inexperienced workers. The organization should step in the process of above matters at which the local authority has to take a serious view and is responsible to reduce the number of material shortages. Respectively department can strengthen the employee's knowledge by providing of more information in the production floor. With FMEA, the organization is able to evaluate processes for possible failures and act as prevention and corrective action rather execute the countermeasure after failures have occurred as FMEA spreadsheet shows in Table 1.

In the FMEA analysis, the risk priority number (RPN) is calculated through the multiplication of ratings given to potential risk occurrence of different hierarchical levels of failure modes. The identification of severity (S), occurrence (O), and detection (D) is based on rating 1 to 10, where rating 1 is a low risk and 10 is a high risk as shows in Table 2.

5   Conclusions

In this paper, a risk assessment for solving the production variation in SMEs has been implemented. Based on the results obtained, the mathematical and statistical analysis formulation for production risk assessment can tackle typical production risks such as retaining backlog information, raw material queuing, product routings, balancing workloads across processes, and preventive maintenance of operating parameters. This study delivers an enhanced approach of risk assessment in production systems. The successful development of production risk assessment will make SMEs to gain competitive advantage in delivering greater value to end customers. The successful development of production risk assessment formulation modelling in this study will ensure the smoothness of the production process and output whereby the developed formulation will be implemented in SMEs.

Table 1 FMEA results

Key process	Failure mode	Failure effects	SEV	OCC	DET	RPN	Action	SEV	OCC	DET	RPN
Product recall	Bottle cap leaking	Food contaminations	8	4	6	192	Traceability test	8	1	6	48
	Metal detection	Food poisoning	9	3	8	216	Traceability test	9	1	8	72
Inappropriate triage process	Misinterpretation	Unreliable result	8	2	4	64					
	Wrong direction	Loss of confidence	8	1	3	24					
Industrial accidents	Human factor	Physical injuries	10	7	6	420	Point deduction of each misconduct	10	2	4	40
	Bad layout		10	3	3	90					
	Internal transportation		8	4	6	192	Pre-test forklift	8	2	5	80

Table 2 Rating scale for severity (S), occurrence (O), and detection (D)

Severity rating	Description	Definition
10	Dangerously high	Death or severe injury to the user
8	Very high	Final result unusable
6	Moderate	Partial breakdown on the result
4	Very low	Performance lost on service parameter
2	Very minor	Minor consequences on performance
1	None	Not notice by the user
Occurrence rating	Description	Definition
10	Very frequent	Error al lease once a month
8	Frequent	Error five to ten times in a year
6	Infrequent	Error once to five times in a year
4	Low	Error once a year
2	Very remote	Error once every five years
1	Remote	Error once a every more than five years
Detection rating	Description	Definition
10	Very remote	Error detection in less than 50% of inspection
7	Very low	Error detection about 50% of inspection
4	Moderately high	Error detection 50% to 70% of inspection
1	Almost certain	Error detection more than 90% of inspection

Acknowledgements This paper is fully supported by finance from Universiti Kuala Lumpur (UniKL) and STRG grant (UniKL/CoRI/str20014). Appreciation is also extended to the selected case study industry and anonymous reviewers for the comments given which led to the significantly improved manuscript quality.

References

1. A.S.Z. Abidin, R.M. Yusuff, R. Muslimen, Exploratory study: design capabilities development for Malaysian vendors in automotive industry. Proc. Int. Conf. Ind. Eng. Oper. Manage. **1**(1), 54–61 (2011)
2. M.N. Che-Ani, A.S. Sapian, I.A. Azid, S. Kamaruddin, Solving production processes disparity issue through implementation of Poka-Yoke concept. IJMMM **5**(4), 278–281 (2017)
3. F. Franceschini, M. Galetto, A new approach for evaluation of risk priorities of failure modes in FMEA. Int. J. Prod. Res. **39**(13), 2991–3002 (2001)
4. P. Koomsap, T. Charoenchokdilok, Improving risk assessment for customer-oriented FMEA. Total Qual. Manag. Excell. **29**(13–14), 1563–1579 (2018)
5. H.C. Liu, J.X. You, X.F. Ding, Q. Su, Improving risk evaluation in FMEA with a hybrid multiple criteria decision making method. Int. J. Qual. Reliab. Manag. **32**(7), 763–782 (2015)
6. H.C. Liu, L. Liu, N. Liu, Risk evaluation approaches in failure mode and effects analysis: a literature review. Expert Syst. Appl. **40**(2), 828–838 (2013)

7. D. P. Möller, Digital manufacturing/industry 4.0. in *Guide to Computing Fundamentals in Cyber-Physical Systems* (Computer Communication Networks, Springer, Cham, 2016).
8. K. Przystupa, An attempt to use FMEA method for an approximate reliability assessment of machinery. ITM Web. Conf. EDP Sci. **15**(1), 05001 (2017)
9. H. Shirouyehzad, R. Dabestani, M. Badakhshian, The FMEA approach to identification of critical failure factors in ERP implementation. Int. Bus. Res. **4**(3), 254–263 (2011)
10. J. Tixier, G. Dusserre, O. Salvi, D. Gaston, Review of 62 risk analysis methodologies of industrial plants. J. Loss Prev. Process. Ind. J. Loss Prevent Proc. **15**(4), 291–303 (2002)

Formulating a Risk Assessment Study for a Job-Shop Production System

Wan Abdullah Wan Sanusi, Mohd Norzaimi Che Ani, and Ishak Abdul Azid

Abstract Risk assessment is a technique used to manage the hazards and risk that potentially cause a harm or injury, especially in workplace. This technique also identifies an appropriate way to eliminate or control the hazard risk. Every workplace has hazards but due to lack of knowledge, underestimated risks, and potentially meeting unexpected situations, are causing the workplace accident. Thus, in this research, a risk assessment has been conducted for the hazard identification and the countermeasure was recommended. The selected case study consists of the job-shop production system, involving the process of regular services and new installation on the customer's site of fire extinguishers. Potential hazards such as exposer to the chemical powder and gases, leakages, and injuries during installation were identified. The theory of Cronbach's alpha has been used in the determination of the worker's awareness by using constructive questionnaires results from selected workers. The obtained results of the developed questionnaires potentially provide a significant result based on the calculated mean score of Cronbach's alpha theory. At the end of this research, a risk control was suggested based on the hierarchy of control to increase safety awareness in the workplace.

Keywords Risk assessment · Hazards · Risk control · Job-shop production system · Safety awareness

W. A. W. Sanusi · M. N. C. Ani (✉)
Manufacturing Section, Universiti Kuala Lumpur, Malaysian Spanish Institute Kulim Hi-Tech Park, 09000 Kulim, Kedah, Malaysia
e-mail: mnorzaimi@unikl.edu.my

W. A. W. Sanusi
e-mail: abdullah.sanusi@s.unikl.edu.my

I. A. Azid
Engineering Section, Universiti Kuala Lumpur, Malaysian Spanish Institute Kulim Hi-Tech Park, 09000 Kulim, Kedah, Malaysia
e-mail: ishak.abdulazid@unikl.edu.my

1 Introduction

Risk assessment is one of the methods used to reduce accident rates in the industry. This method describes the management of hazards and risk factors that potentially cause harm or injury in the workplace. It also determines appropriate ways to eliminate the potential hazards and control the identified risk. Normally, hazard identification, risk assessment, and risk control (HIRARC) methodology has been used in conducting the risk assessment for any workplace. HIRARC is a universal method widely utilized by occupational safety and health (OSH) practitioners in any field of industry and it is one of the compulsory items as prescribed under the Occupational Safety and Health Act 1994 [1]. In addition, the importance of HIRARC is to identify the potential hazard within the workplace and responsibility toward laws that may relate to ISO45001:2018. The organization organizing any procedures could be to ensure the identification of risk, risk assessment, and its determination. The method of HIRARC is often stated as the identification or classification of all health hazards. This quantitative assessment of risk for all uncertainty hazards and control of unacceptable health risks was implemented in the selected case study organization.

Identification of hazards means that unwanted occurrences leading to the materialization of the risk and the mechanism by which such unwanted events may occur are determined [2]. The objective of hazard identification is to identify key tasks, such as jobs that pose a significant danger to employee health and safety, and to identify dangers connected with equipment related to energy, working circumstances, and activities. Three key classes, health threats, safety hazards, and environmental hazards, can be categorized into hazards [3]. A safety threat is any force that is powerful enough to cause harm or injury while performing routine tasks. Normally, the workplace accident caused by a safety threat from unwanted occurrences. A worker can be badly cut, for instance. When workplace controls are not sufficient, safety hazards cause damage.

Risk assessment is an entire process by which the risk magnitude is estimated and whether the risk is bearable or not. A risk can be defined in various ways to communicate analytical data to make risk management decisions. The outcome in a risk matrix is an extremely efficient approach in solving the safety issues in workplace for risk analysis which uses probabilism and seriousness in qualitative technique. This method is regularly applied by established companies or those companies certified in ISO45001:2018. Unfortunately, less knowledge of risk assessment application is available in small medium enterprises (SMEs) or micro-enterprises, especially those operating their business using job-shop production system. The data of the injury might be hidden or unreported, but a safety awareness program must be applied to that type of organizations.

Thus, this study is focused on the development of constructive questionnaires toward increasing the awareness of workplace hazards in the selected case study industry. The developed questionnaires are applied systematically to identify the awareness level of safety and health in the workplace. The selected case study industry is specialized in fire protection systems which consist of installations related to high-rise buildings, factories, housing projects, hotels, resorts, and other heavy industries. The scopes of this study were focusing on services and maintenance, installation of new fire extinguishers on customer's sites. The next section of this paper discusses the literature review, followed by the descriptions of the risk assessment method. Later, the application of the risk assessment in the selected case study is discussed and, the result and discussion are also highlighted. Finally, the finding will be concluded at the end of this paper.

2   Literature Review

This section discusses the related information regarding the literature review focusing to this study. The review of the published literatures was performed based on obtained resources through online databases such as journals, reference books, case studies, and other sources associated to occupational safety and health (OSH) in general and the specification of risk management. All information that is related to the research that has been deliberated in this section then will be applied in the methodology and discussion of the research to ensure the defined objectives are achievable.

Occupational Safety and Health Act (OSHA) 1994 is an act that provides a regulatory basis for stimulating and enforcing high safety and health standards at work. OSHA provides guidance to carry out the routine duties of employers, workers, manufacturers, and suppliers. It also allows for the introduction of codes of conduct in the industry and the making of regulations. The application of the act to "persons at work" is a significant feature of OSHA that departed from the earlier law form [4]. The main objectives of this act are to secure the health, safety, and welfare of persons against risks from work activity. The second objective is to promote the safer workplace that is adapted to physiological and psychological needs. The last objective is to provide the conducive working environment synchronized to a system of regulations and approved industry codes of practice.

HIRARC is one of the general tasks required by OSHA 1994 for the employer to provide its workers with a secure workplace [5]. Three consecutive phases of HIRARC are hazard identification (HI) which consists of identification of hazards implies the identification of undesired events leading to the hazard's materialization [6]. The objective of hazard identification is to identify key tasks, such as those jobs which represent considerable risks to the health and safety of employees [7]. Three key classes of hazards are known as safety environmental hazards, hazards, health hazards, and hazard identification techniques. The second phase is known as risk assessment (RA), which consists of identifying the possibility of injury or harm occurring to a worker if exposed to a hazard. A hazard is considered an agent that possibly affects the humans, property, or the environment. The last phase is risk control (RC), which is the step of measures at which potential hazards will be eliminated or reduced by the risk of a person being exposed to a hazard. HIRARC will help to create corrective actions that may be performed to reduce the possibility for exposure to a hazard, remove the danger, or reduce the likelihood of recognizing the risk of exposure to that hazard. The safe protection of moving elements of equipment will be a basic control method that will eliminate the possibility of contact. When we look at control measures, we also consider the hierarchy of control measures.

Small and medium enterprises (SMEs) play an important role in the economic growth in Malaysia. SMEs are important to sustain the economic development and public stability, creating employment opportunities, perfecting market economic system production and operation ability in modern market economy competing more sharply and the environment of global economic integration [8]. SMEs need to alert the challenges of the production and operation capacity and capability and then promptly execute the action plans toward effective management tactic according to its production and operation ability. Unfortunately, according to an annual report from the Social Security Organization (SOCSO) between the years 2010 and 2015, SMEs in the field of the metal industry have the highest reported number of workplace accidents compared to the other fields [9]. This indicator described the lack of knowledge and ignoring the hazards as potentially the main causes of this situation. In addition, based on the results of review from published works of literature, the HIRARC application in SMEs shows that most of the employees are unfamiliar with that concept, and this situation easily exposes themselves to workplace accidents [10]. Therefore, the risk assessment of operation ability in SMEs is very important. Hence, it is indeed important to increase awareness among the employees and employer of the SMEs to promote and emphasize the safety in workplace.

3 Methodology

The research methodology in this paper consists of two main phases known as development of the constructive questionnaires and risk assessment using the HIRARC model as illustrated in Fig. 1. In the phase 1, the questionnaire has been developed to identify the safety awareness among the workers in the selected case study industry divided into two main sections which are part A and part B, while, in the phase 2, the common HIRARC model has been applied to identify the potential risks in the selected case study industry.

3.1 Development of Questionnaire

This research questionnaire is based on a structured type that has been used. The questionnaire is divided into two main parts. Part A is regarding the personal particulars of the participants. Part B is divided into six elements of the area that need to be analyzed as summarized in Table 1. Multiple-choice answers consist of six points of answers that have been applied. The six-points answers are adapted from the Likert scale as shown in Table 2. The Likert scale is one of the rating scales that

Fig. 1 Research methodology

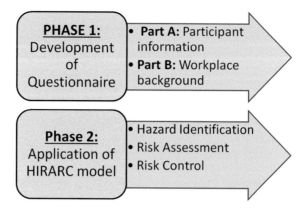

Table 1 Element in questionnaire

Part	Contents
A	1. Gender
	2. Position
	3. Age
	4. Working experience
	5. Process involved
B	1. Working environment
	2. Facilities and equipment
	3. Personal protective equipment
	4. Information procedures
	5. Work-in-progress

Table 2 Six points of Likert scale

Likert scale	Description
1	Strongly disagree
2	Disagree
3	Slightly disagree
4	Slightly agree
5	Agree
6	Strongly agree

has been widely used to measure attitudes directly. This scale is also known as a six-point scale which is used to allow the respondents to state their level of agreement or disagreement with a particular statement. The Likert-type scale is used in determining the level of satisfaction of the employees upon the services provided by the chosen company. The selected respondents will be answered by using the 6-points Likert scale method, and then the results will be analyzed to ensure reliability test.

Table 3 Value of Cronbach's alpha

Cronbach's alpha	Internal consistency
$\alpha \geq 0.9$	Excellent
$0.9 > \alpha \geq 0.8$	Good
$0.8 > \alpha \geq 0.7$	Acceptable
$0.7 > \alpha \geq 0.6$	Questionable
$0.6 > \alpha \geq 0.5$	Poor
$0.5 > \alpha$	Unacceptable

A pilot test has been conducted to study the reliability of the developed questionnaire based on the answers from selected workers. The feedback of the respondents is used to analyze the structure and the content of the questionnaire. The validity test or face validity test has been applied to measure the level of the questionnaire. A reliability test is the measurement that shows the stability and consistency of a measuring instrument. The internal consistency is measured using Cronbach's alpha, which is the most widely used method as shown in Table 3.

3.2 Application of HIRARC Model

Application of the HIRARC model requires judging the index of likelihood (L) and severity (S) by using a scale of 1 to 5. Scale 1 represents the lowest risks and scale 5 represents the highest risks. In the determination of the L and S scale, normally the mutual agreement between selected team members is required. Since the scale has been used in the determination of the risk level, the references of each scale have been classified as shown in Tables 4 and 5.

Once the index scale has been identified for L and S, the risk index (RI) will be calculated by using Eq. (1).

$$\text{Risk index}(\text{RI}) = \text{Likelihood}(\text{L}) \times \text{Severity}(\text{S}) \tag{1}$$

Once the RI has been identified, the overall risk index will be evaluated based on three classes which are high, medium, and low. In the determination of the class as

Table 4 Likelihood (L) rating scale

Likelihood (L)	Description	Scale
Most likely	Hazard will be experienced and realized	5
Frequent	Hazard will be occurring not usual	4
Probable	Good chance of occurring	3
Remote	Manifestations of the hazard are possible but unlikely	2
Improbable	Practically impossible and never occurred	1

Table 5 Severity (S) rating scale

Severity (S)	Description	Scale
Fatal	Damage of human body or property	5
Serious	Permanent injury/disability, severe injury	4
Major	Occupational disease/poison/chemicals	3
Minor	Minor abrasions, bruises, cut, first aid types injured	2
Near miss	No significance risk of injury	1

Table 6 Indication of the risk matrix table

Likelihood (L)	Severity (S)				
	1	2	3	4	5
5	5	10	15	20	25
4	4	8	12	16	20
3	3	6	9	12	15
2	2	4	6	8	10
1	1	2	3	4	5

per the obtained RI score, then it will be evaluated to classify the level or risks by referring to Tables 6 and 7.

Finally, the hazard should be controlled at its source and the closer it gets is better. There are also several steps needed to be done in order to achieve the best-determining control.

4 Result and Discussion

The selected case study consists of the job-shop production system, involving the process of regular services and new installation on the customer's site of fire extinguishers. Potential hazards such as exposure to the chemical powder and gases, leakages, and injuries during installation were identified.

4.1 Result of Questionnaire

The questionnaire consists of two (part A and part B) and has been successfully distributed to the workers of the selected company. The results obtained from the questionnaire are tabulated in Table 8. Table 8 shows the results from the respondents

Table 7 Determination of priority based on ranges

Risk	Description	Action
15-25	High	A ranking of the highest risk requires immediate corrective action to control the hazard. Actions taken must be proper documented on the registered form including all required information such as date, signature, person in-charge etc.
5-12	Medium	A ranking of medium risk requires a proper planning approach in controlling the identified hazards and temporary action must be applied if required. Actions taken must be proper documented on the registered form including all required information such as date, signature, person in-charge etc.
1-4	Low	The acceptable ranking because of the low risks and further reduction may not be necessary. However, if the risk possible to resolved quickly and efficiently, all the action plans should be implemented and recorded.

in the demographic information. Most of the respondents are male which is 93.33%. Most respondents are workers by 90%. 13 of the employees are in the range from 18 to 25 years old in age contributing by 43.33% of the respondents. 11 respondents have a one- to three-year work experience, and the percentage is 36.67%. 6 of 30 respondents have worked as installing a new fire extinguisher process worker and contributed 20.00%.

The developed questionnaire shows the significant result where the overall results of the Cronbach alpha are classified in the "acceptable" class (as mentioned in Table 3). The overall alpha value from the questionnaire results was 0.7477. Among the elements, facilities and equipment were perceived as rather low with a mean score of 4.771 (Table 9).

4.2 Result of Implementation of the HIRARC Model

HIRARC's objective is to discover any elements that may endanger employees and others. Table 10 represents the result of risk assessment by using the HIRARC model for determining the possible hazard, severity, probability, and risk of each procedure in the selected case study industry. The recommended countermeasures have been suggested at the end of Table 10.

Table 8 Result of questionnaire

Item	N (30)	Cronbach alpha, α (%)	Item	N (30)	(%)
Gender			*Position*		
Male	28	93.33	Manager	1	3.33
Female	2	6.67	Supervisor	2	6.67
Age			Worker	27	90.00
			Working experience		
			< 1 years	4	13.33
< 18 years	0	0.00	1–3 years	11	36.67
18–25 years	13	43.33	4–6 years	8	26.67
26–35 years	9	30.00	7–10 years	3	10.00
36–45 years	6	20.00	11–15 years	3	10.00
46–60 years	2	6.67	16–20 years	0	0.00
> 60 years	0	0.00	20–30 years	1	3.33
			> 30 years	0	0.00
Process involved					
1. Refill chemical powder process	2	6.67			
2. Refill gas process	2	6.67			
3. Wash and clean process	2	6.67			
4. Maintenance process	3	10.00			
5. Weighing process	3	10.00			
6. Unloading process	5	16.67			
7. Data marking process	2	6.67			
8. Install a new fire extinguisher process	6	20.00			
9. QC visual inspection	2	6.67			

Table 9 Result of Cronbach alpha, α

Variables	Mean	Cronbach alpha
Work environment	4.856	0.3604
Facilities and equipment	4.771	0.6924
Personal protective equipment	4.967	0.6924
Training and supervision	5.120	0.7392
Information procedure	5.042	0.4097
Work process	5.150	0.5321
Overall		0.7477

Table 10 Results of risk assessment using HIRARC method

	Hazard Identification			Risk Analysis					Risk Control
	Work Activity	Hazard	Effect	Existing risk control (if any)	L	S	RI		Control measure
1	Unloading and weighing the fire extinguisher	Sharp edge	Cut / first aid types of injury	Safe work practice	3	1	3	L	Use leather hand glove
		Heavy load	Muscular strain / back pain	Manual lifting procedure	3	3	9	M	Use carts
2	Refill chemical powder and gas into the fire extinguisher	Inhale chemical powder	eyesight	Wear mask	3	3	9	M	Wear protection and goggle
3	New installation	Crushed foot / hand stuck	Foot and hand injury	Safe work procedure	4	4	16	H	Warning signboard and wear PPEs
4	Washing and cleaning the fire extinguisher	Slippery surface	Slippery when perform the task	Wear safety boots	3	3	9	M	Safe work procedure, use the PPEs
5	Maintenance and drying using air gun	Noise	Ear damage	N/A	3	3	9	M	Wear earmuffs and earplugs
6	QC inspection	Hot	Dehydrate	N/A	4	3	12	M	Conducive workplace
7	Data marking	Exposure to the computer screen	Eyestrain	N/A	4	3	12	M	wear glasses

5 Conclusion

This study was set out to study the hazard identification, risk assessment, and risk control (HIRARC) at a selected case study industry, which is specifically a fire extinguisher company. The hazard identification starts with the observation checklist and comes out with the questionnaire based on the checklist. The risk assessment is based on the observation at the company and the distribution of the questionnaire to employees. After being analyzed, the result of the overall questionnaire on Cronbach's alpha value is 0.7477, which is acceptable based on the Cronbach alpha (α) value. The obtained results of the developed questionnaires potentially provide a significant result based on the calculated mean score of Cronbach's alpha theory. At the end of this research, risk control was suggested based on the hierarchy of control to increase the safety awareness in the workplace. In order to identify the control measure, the observation has been made and identified the existing control measure and recommended new control measure from the feedback given in the questionnaire. From the HIRARC form that has been fulfilled, one of the processes was at high risk which is installing the new fire extinguisher process. Effect on the foot and hand injury can cause fatal injury. Immediate action to control the hazard as recommended in the HIRARC form should be applied. Six of the processes are at medium risk, and one is at low risk. The focus of this study is primarily on improving safety and health. Future research should focus on additional ways to improve the workplace and create a better working environment.

Acknowledgements This paper is fully supported by finance from Universiti Kuala Lumpur (UniKL) and STRG grant (UniKL/CoRI/str20014). Appreciation is also extended to the selected case study industry and anonymous reviewers for the comments given which led to the significantly improved manuscript quality.

References

1. J. Lee, J. Jung, S.J. Yoon, S.H. Byeon, Implementation of ISO45001 considering strengthened demands for OHSMS in South Korea: based on comparing surveys conducted in 2004 and 2018. Saf. Health Work **11**(4), 418–424 (2020)
2. B.A. Aubert, M. Patry, S. Rivard, A framework for information technology outsourcing risk management. ACM SIGMIS Database Data Base Adv. Inf. Syst. **36**(4), 9–28 (2005)
3. D. Lithner, A. Larsson, G. Dave, Environmental and health hazard ranking, and assessment of plastic polymers based on chemical composition. Sci. Total. Environ. **409**(18), 3309–3324 (2011)
4. A. Leino, Intranet-based safety documentation in management of major hazards and occupational health and safety. Int. J. Occup. Saf. Ergon. **8**(3), 331–338 (2002)
5. S.B.M. Tamrin, I.M. Yusoff, 1 hazards in workplace. Occupational safety and health in commodity agriculture: case studies from Malaysian agricultural perspective **1**(1), 55–72. Selangor, Malaysia (2014)
6. V. How, K. Karuppiah, Filling the gaps of the workplace first aid assessment by considering the guidelines on occupational safety and health Malaysia. APEOHJ **1**(1), 23–28 (2015)
7. M.A. Hamka, Safety risks assessment on container terminal using hazard identification and risk assessment and fault tree analysis methods. Procedia. Eng. **194**(1), 307–314 (2017)
8. G.G. Fiseha, A.A. Oyelana, An assessment of the roles of small and medium enterprises (SMEs) in the local economic development (LED) in South Africa. Am. J. Econ. **6**(3), 280–290 (2015)

9. M.Z.M. Saat, C. Subramaniam, F.M. Shamsudin, A proposed relationship between organizational safety practices and safety performance in the manufacturing of small and medium enterprises in Malaysia. Sains Humanika **8**(4–2), 91–97 (2016)
10. V.Y. Siong, J. Azlis-Sani, N.H.M. Nor, M.N.A.M. Yunos, J.A. Boudeville, S. Ismail, Ergonomic assessment in small and medium enterprises (SMEs). J. Phys. Conf. Ser. **1049**(1), 012065 (2018)

Effective Safety and Health Management System Model to Improve in Malaysian Small, Micro and Medium Enterprises (SMEs)

Y. Shahrizan, Mohamad Alif, S. M. Syed Ahmad Faiz, M. S. Sazali, and M. K. Huda

Abstract There has been a decline and the number of injuries in the workplace in recent times. However, micro, small and medium enterprises remain one of the high-risk industries. The purpose of this study is to explore the important of safety and health, to develop a model to improve the implementation of safety and health in micro, small and medium enterprises in Malaysia and also to verify the developed safety and health management system model using survey. This industry has made a significant contribution to the country's economic growth. However, when safety and health management is not implemented systematically, accidents can occur and this can affect the country's economic growth.

Keywords Project management · Safety and health · Malaysian SMEs

Y. Shahrizan (✉) · M. Alif · S. M. Syed Ahmad Faiz · M. S. Sazali · M. K. Huda
Universiti Kuala Lumpur, Malaysian Spanish Institute Kulim Hi-Tech Park, 09000 Kulim, Kedah, Malaysia
e-mail: mshahrizan@unikl.edu.my

M. Alif
e-mail: Alif@s.unikl.edu.my

S. M. Syed Ahmad Faiz
e-mail: syedahmadfaiz@unikl.edu.my

M. S. Sazali
e-mail: msazali@unikl.edu.my

M. K. Huda
e-mail: mhuda@unikl.edu.my

M. H. Abu Bakar et al. (eds.), *IT Solutions for Sustainable Living*,
SpringerBriefs in Applied Sciences and Technology,
https://doi.org/10.1007/978-3-031-51859-1_5

1   Introduction

The small, micro and medium enterprises (SMMEs) sector is as an economic engine for many countries. These small, micro and medium enterprises have played significant role in world economies in terms of economic growth, employment and stimulating investments. Researchers in the field of entrepreneurship agree that this sector is crucial for economic growth, employment creation, poverty reduction and reducing levels of inequality. This sector is very crucial and plays a significant role in marginalized and rural provinces. In Malaysia, SMMEs also continue to contribute to the economy as Malaysia embarks on the journey toward achieving Vision 2020.

A safety and health management system (SHMS) is a systematic approach put in place by an employer to minimize the risk of injury and illness. It involves identifying, assessing and controlling risks to workers in all workplace operations. One of the most effective ways to reduce workplace hazards and injuries is through a comprehensive, proactive safety and health management system. The benefits of implementing safety and health management systems include protecting workers, saving money and making all your hazard-specific programs more effective.

However, small, micro and medium enterprises lack knowledge, skills and other resources to implement OSH in their workplace. Majority of micro, small and medium enterprises face difficulty in implementing OSH as they lack expertise, resources or manpower. Besides, they also lack of financial resources and expertise clearly impedes their ability to implement OSH effectively. This project is to identify the cause of hazard, explore the important safety and health, to develop a model to improve the implementation of safety and health among micro, small and medium enterprises and to verify the developed safety and health management system model using survey. It can be concluded that the safe workplace environment will eliminate the workers pain and suffering, reduce absences and increase productivity, increase motivation and the commitment of employees and reduce business costs, achieve compliance to legislation.

Thus, the objective of this project is to study the safety and health management for small, micro and medium enterprises. The study is to explore the important of safety and health among Malaysian Small, Micro and Medium Enterprises. Besides, this study is to develop a model to improve the implementation of safety and health among small, micro and medium enterprises.

2   Literature

This chapter will cover the research of safety and health management for small, micro and medium enterprises. Their topic and findings are closely related to this study. The main purpose of this chapter is to summarize the safety and health management for small, micro and medium enterprises (SMMEs).

The role of safety training in promoting safety behaviors among workers has been widely documented [1]. When effective, safety training improves employee safety, skills and knowledge. They tend to be aware of potential hazards and risks in the workplace and the possible consequences accident if they are not followed. While formal security programs tend to have syllabuses and structured activities for employees to engage in, their implementation may be over-budgeting SMMEs. It has been said that SME employers prefer to dedicate their resources to other things of more productive value [3]. In the context of SMEs, when sending workers to formal security programs or running internal and structured security programs may not be the most viable option due to resource constraints, informal and less structured training can be performed daily. For example, SMEs employers can set aside time for informal training sessions before the start of the working day on security issues. However, this assumes that SMEs employers themselves have undergone several formal security programs.

Workplace safety is an important issue in many organizations. This is because accidents and injuries in the workplace can bear organizational costs both financially and non-financially. Although safety at work is less checked, less attention is given to small, micro and medium enterprises. Such neglect will cause many accidents and injuries worldwide, including Malaysia, in this organization.

Safety and feedback communication has been recognized as an effective way to improve safety performance in organizations [2]. The dissemination of information through various communication media, such as safety meetings, organized personal contacts, and signage and others on safety regulations and laws can serve as a reminder to employees about the need to keep safe and work safely [4]. However, to be effective, safety and feedback communication must be a two-way process and not just a top-down approach. Employees should also be encouraged to provide their feedback on safety related matters to management and suggest ways to improve work processes and activities that can be made safer (Fig. 1).

3   Methodology

Surveys, questionnaires and interviews are the common tools of research. The research activity starts with literature review survey by finding and collection of related journals and articles on Evm whether it apply on the industry. The conceptual model will be created based on the findings of literature review.

3.1   Design of Conceptual Model

In the analysis, the conceptual model is based on the results of the previous study. The conceptual model is created from the previous study that will influence the outcome of the study. The study shows that the model of the safety and health

Fig. 1 Personal protective
equipment

m management system has three key findings. The key results are management commitment, employee commitment, and the last one is the advantages of OSH implementation. There are many types of conceptual model. The conceptual models that can be adopted in this study (Fig. 2).

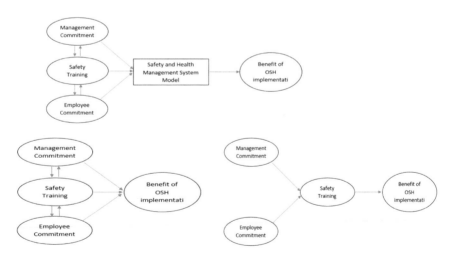

Fig. 2 Conceptual model

3.2 Distribute Questionnaire

Questionnaires were distributed to managers and employees in several factories around Kulim and Penang in the manufacturing, services and other sectors of industry and have received feedback. Since this study is a quantitative study, at least 80 feedbacks are required. The questionnaire consists of four parts that is background of enterprise, the role of safety and health management, safety and health training, the role of employees on safety and health and the benefits of OSH implementation.

4 Result and Discussion

Table 1 until 4 show the result from the perspective of management commitment, we can conclude that many parties are beginning to realize the importance of manager commitment in the implementation of safety and health management system.

Table 2 from the results of the table survey, many people argue that safety training is important to increase knowledge about safety in the workplace.

Table 1 Descriptive statistics

	N	Minimum	Maximum	Mean	Std. deviation
Top management involved in safety and health management aspects?	85	1	5	3.53	1.230
Worksite policy provides an understanding of the commitment and expectations of management to have a safe and healthy work environment?	85	1	5	3.65	1.222
Have a system for identifying, preventing and dealing with hazards?	85	1	5	3.71	1.326
Managers responsible for safety and health system?	85	1	5	3.66	1.268
Do the organization conducts annual reviews	85	1	5	3.69	1.235
Valid N (listwise)	85				

Table 2 Descriptive statistics

	N	Minimum	Maximum	Mean	Std. deviation
Everyone gets health and safety training at work?	85	1	5	3.52	1.419
Does safety training enable workers to gain more knowledge on safety at a faster rate?	85	1	5	3.59	1.348
Do employees will be exposed to safety issues that affect them in their workplace?	85	1	5	3.71	1.242
Does safety training improve risk perception of employees?	85	1	5	3.86	1.274
Does safety training help to reduce accident rate?	85	1	5	3.78	1.357
Valid N (listwise)	85				

Table 3 many admit that workers often play a role in this area of protection because they are vulnerable to danger from the results of the table survey.

Table 4 shows the result of benefits of OSH implementation. Many admit implementation OSH has many of benefits such as eliminating the workers pain and suffering, reducing absences and increasing productivity, increasing motivation and the commitment of employees and reducing business costs.

4.1 Conceptual Model

The conceptual model on Fig. 3 shows management commitment, safety training and employee commitment are contributing to the benefits of OSH implementation but employee commitment are more contribute more percentages of benefits of OSH implementation. This is because the employees are more face the hazards in the company. However, safety training is also important to employees to add knowledge about hazards and accidents. Management also needs to play a role by monitoring safety aspects from time to time.

5 Conclusions

There are still some micro, small and medium companies that underestimate the safety of employees. This is because they do not face major problems such as accidents involving deaths and assets. Therefore, they feel comfortable with the current situation. In addition, the employer does not emphasize the safety and health of employees. For example, employees do not wear PPE while doing work due to low safety awareness. Other factors that may cause this low safety awareness are due to

Table 3 Descriptive statistics

	N	Minimum	Maximum	Mean	Std. deviation
There is an active and effective health and safety committee	85	1	5	3.48	1.306
Responsible to ensure that workplace is neat and safe	85	1	5	3.75	1.253
Report accidents or anything else they think could be unsafe or unhealthy	85	1	5	3.78	1.267
Communication on safety and health procedures is done in a way that employees can understand?	85	1	5	3.74	1.207
There is regular communication between employees and management about safety issues?	85	1	5	3.72	1.269
Incidents and accidents are investigated quickly in order to improve workplace health and safety?	85	1	5	3.71	1.252
Valid N (listwise)	85				

the cost of PPE payment, safety and health of training officers and salaries. This is a huge investment for small businesses.

Table 4 Descriptive statistics

	N	Minimum	Maximum	Mean	Std. deviation
Protects and enhances an organization's reputation and credibility	85	1	5	3.74	1.264
Helps maximize the performance and/or productivity of employees	85	1	5	3.68	1.338
Eliminates or minimizes risks to employees	85	1	5	3.76	1.260
Reduces business costs and disruption	85	1	5	3.71	1.252
Shows evidence of systematic arrangements are in place to carry out the OSH policy and objective	85	1	5	3.84	1.194
Enhances employees commitment to the organization as a whole	85	1	5	3.69	1.345
Improves employee attitude toward health and safety	85	1	5	3.80	1.213
Improves conformance with legal requirements	85	1	5	3.74	1.302
Continually improves OSH performance	85	1	5	3.81	1.277
Valid N (listwise)	85				

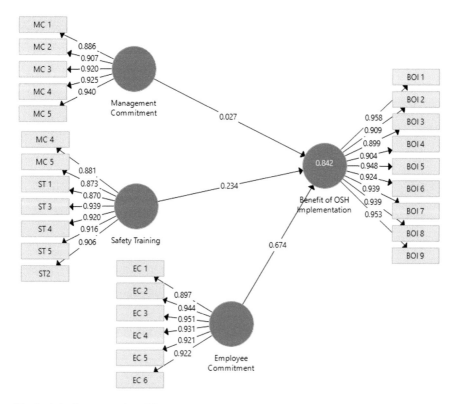

Fig. 3 Actual conceptual model

References

1. L. Surienty, K. Hong, D. Kee, Occupational safety and health (OSH) in Malaysian small and medium enterprise (SME) and effective safety management practices. Int. J. Bus. Technopreneurship **1**, 321–338 (2011)
2. C. Subramaniama, F.M. Shamsudinb, M. Lazim, M. Zina, S.S. Ramalub, Z. Hassana, Safety management practices and safety compliance: a model for SMEs in Malaysia. Eur. Proc. Soc. Behav. Sci. (2016)
3. A.A. Aziz, M.E. Baruji, M.S. Abdullah, N.F.N Him, N.M. Yusof, An Initial study on accident rate in the workplace through occupational safety and health management in sewerage services. Int. J. Bus. Soc. Sci. **6**(2) (2015)
4. L. Surienty, K.T. Hong, D. Kee, Occupational safety and health (OSH) in SMES in Malaysia: a preliminary investigation. J. Glob. Entrepreneurship (2011)

Improving the Housing and Building Maintenance Cost Effectiveness and Purchasing Strategy Using FMEA and Life-Cycle-Cost Measurement

Fauzan Rahman, Mohamad Sazali Said, Azmi Hassan, Mohd Shahrizan Yusoff, Surya Atmadyaya, and Yusri Yamin

Abstract Cost-effectiveness analysis plays a crucial role in the purchasing Strategy. In order to optimize the cost benefits, analysis of failure data is essential. This research proposes the purchasing strategy of the tube lamp model using life cycle cost (LCC). The study combined failure modes and effect analysis (FMEA) method and the LCC analysis for a better decision making in the order to improve the maintainability, maintenance effectiveness and reduce the maintenance costs. Qualitative and quantitative analysis helps to identify critical parameters and recommended maintenance action. It also includes how cost–benefit analysis is useful to evaluate the life cycle costs of the measured object.

Keywords Life cycle cost (LCC) · Failure modes and effect analysis (FMEA) · Tube lamp · Purchasing · Reliability analysis

---

F. Rahman (✉) · M. S. Said · M. S. Yusoff · S. Atmadyaya · Y. Yamin
Manufacturing Section, Universiti Kuala Lumpur, Malaysian Spanish Institute Kulim Hi-Tech Park, 09000 Kulim, Kedah, Malaysia
e-mail: fauzan.rahman@s.unikl.edu.my

M. S. Said
e-mail: msazali@unikl.edu.my

M. S. Yusoff
e-mail: mshahrizan@unikl.edu.my

S. Atmadyaya
e-mail: atmadyaya.surya@s.unikl.edu.my

Y. Yamin
e-mail: yamin.yusri@s.unikl.edu.my

A. Hassan
Electrical, Electronics and Automation Section, Universiti Kuala Lumpur, Malaysian Spanish Institute Kulim Hi-Tech Park, 09000 Kulim, Kedah, Malaysia
e-mail: azmi.hassan@unikl.edu.my

© The Author(s), under exclusive license to Springer Nature Switzerland AG 2024
M. H. Abu Bakar et al. (eds.), *IT Solutions for Sustainable Living*,
SpringerBriefs in Applied Sciences and Technology,
https://doi.org/10.1007/978-3-031-51859-1_6

1   Introduction

In a manufacturing company, the house and building maintenance is categorized as non-manufacturing costs. Every enterprise is asked to have optimum cost benefits, and in more detail, several enterprises put the cost efficiency as their key performance indicator (KPI). When the actual operational budget forces to significantly cut off their total cost, likely a company starts from the non-manufacturing costs. Thus, the cost of equipment procurement is highly important to consider carefully. The price is the main factor when deciding the suppliers and this is a common phenomenon that happens in actual purchasing activity [1]. However, the cost of equipment must be assessed over the course of its full lifetime, including equipment failure, maintenance fees, and operational costs, rather than relying solely on the initial investment. In the total LCC, the cost of unplanned breakdown and production loss is a significant issue of considerations [2]. A product or equipment life cycle costs can be multiple times compared to the initial purchase or early investment costs [3]. As a result, the LCC should be considered in the purchasing strategy to maximize overall performance.

2   Literature

2.1   FMEA

FMEA is a useful tool for ensuring preventive reliability and quality. FMEA is a commonly used tool in manufacturing for risk assessment and quality improvement [4]. In assessing potential risk [5] and other fields, FMEA is a tool that can be used to detect failure modes in a product or design as early as the design phase, by investigating and evaluating potential causes and effects [6]. FMEA was utilized in this study to analyse the equipment failure mode during normal operation and routine maintenance. Only qualitative scales for severity and detection are considered in the FMEA [7]. The ordinal scale variables calculation are using the three indices to obtain risk priority number (RPN) [8]. The three indices that help establish the priority of failures are occurrence (O), severity (S), and detection difficulty (D). The failure frequency is called occurrence, the severity of the risk is called severity, and the detection difficulties are best described as the failure before it reaches the customer.

2.2   Life Cycle Cost

The total "lifetime" of any piece of equipment including the purchasing, installation, operation, maintenance, and disposing costs of that equipment is described as the life cycle cost (LCC). By following the methodology to identify and quantify all of the parameters of the LCC formula/equation, we can get the LCC of the measured

Table 1 Maintenance order category

Division	Count	%	Maint. order	Sum	%
Civil and building	1	25	12	12	7.80
Road	1	25	22	34	14.40
Small repair	1	25	54	88	35.30
Electrical	1	25	65	153	42.50
$N = 4$					

product or equipment. In order to compare the tool or material between current product specifications and proposed alternatives, the LCC is able to deliver which solution is expected as the most cost-effective within the limits of the available data.

In the 1960s, the Department of Defense of the US developed the LCC concept, and in 1971, the LCC was incorporated as policy by Directive 5000.1, Acquisition of Major Defense Systems [2], and later on in the industrial system and consumer product areas. The LCC concept has been used for a wider scope [9]. The economic approach within a standard engineering context is useful to decide between alternative products or designs that provide equal service to the customer [10]. Today, in construction at home, industries, and abroad, the LCC analysis is widely used.

3   Methodology

The risk priority number (RPN) is used to assess the FMEA results; the failure mode of the product defines the risk level. This figure is obtained by multiplying the failure mode indices. In the case of electrical issues, the occurrence of the problems can vary, as shown in Table 2.

- Severity (S).
- Occurrence (0).
- Detection difficulty (D).

$$RPN = S \times O \times D. \tag{1}$$

Table 2 Maintenance activity with electrical category

Problem	Count	No. of problem	Count
Accessories replacement	1	20	1
Broken lamp	1	35	1
Power source loss	1	40	1
Wiring problem	1	167	1
$N = 4$		$N = 4$	

Acquisition, running costs, maintenance costs, cost of recovery measures, and final disposal fees are all included in life cycle costs. The following is the formula:

Costs Involved

- Initial cost (C_{ic}).
- Delivery and installation costs (C_{inst}).
- Energy cost (C_e).
- Operational cost (C_o).
- Maintenance cost (C_m).
- Downtime costs (C_s).
- Decommissioning/disposal costs (C_d).

$$LCC = C_{ic} + C_{inst} + C_e + C_o + C_m + C_s + C_d \qquad (2)$$

In this study, the LCC of a tube light can be divided into two main categories, i.e. initial costs and operational and maintenance costs.

Initial Cost (C_{ic}):

- Cost of one tube.
- Life expectancy of one tube.
- Number of tube lights needed to burn for 15.000 h.
- The result is the total payment to tube light for 15.000 h.

Operational and Maintenance Cost (C_{omc}):

- Hours required to use one kilowatt of electricity for the size watt tube light.
- kWh of electricity in 15.000 h.

Result is the total cost of electricity used.

$$LCC = C_{ic} + C_{omc}. \qquad (3)$$

4   Failure Mode Effect Analysis

The maintenance activity categories were sorted using a Pareto chart to get the dominant cause and then by using FMEA to find the main failure mode from the selected issue by calculating the possibility of the incident and accident. This prioritization enables us to understand which part of the product or process is the most impactful to focus on during design or formulation, to obtain a better acceptance [11] so the new product development or current products improvement can get the benefit [12]. The electrical issue obtained a value of 42.5% from the total maintenance order as shown in Fig. 1 and decided as the maintenance activity category that will be calculated. There are several dominant issues in the electrical issues such as lighting problem,

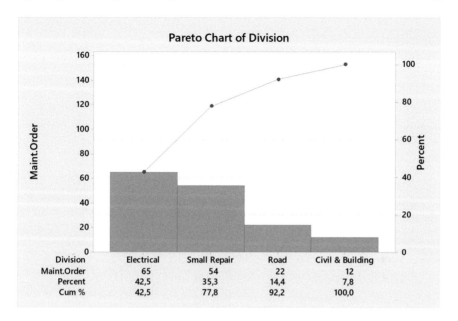

Fig. 1 Pareto chart of maintenance order category

power source problem, office/house electrical wiring problem, and electrical accessories problem which are categorized as the failure mode in the next calculation. Then, the failure causes were investigated by setting up the severity, frequency of occurrence and range of the difficulty. Items with high RPN indices can be identified with high numbers, as shown in Table 3. In this study, the items with a RPN index above 100 are taken as the focused item. As a result, the proposed actions for improvement are suggested. Such as, the cause of the highly lighting problem is the lack of material (lamp) quality. Suggestions were provided, such as the lamp model, type, and specifications selection so it will help the person in charge to plan the stock or recalculate the minimum specifications. The Risk Priority Number (RPN) generated from the Failure Mode and Effects Analysis (FMEA) is used to calculate the Life Cycle Cost (LCC). The measurements for Corrective Inspection Cost (C_{ic}) and Corrective Operation & Maintenance Cost ($Comc$) used in this LCC calculation are relatively static, particularly the assumed electricity usage ratio.

5 Case Study

Pareto diagram (see Fig. 2) for the maintenance categories of the electrical problem was created as the selected issue. Table 3 is the FMEA table for electrical maintenance order. Broken lamp achieved the highest RPN with value of 168 is selected as the

Table 3 FMEA of maintenance activity under electrical category

No	Process map—activity	Key process input	Potential failure mode	Potential failure effects	SEV	Potential causes	OCC	Current controls	DET	RPN	Recommended	Responsibility
1	Lighting	Replace	Broken lamp	No lighting	7	Lack of quality	6	As request	4	168	Lamp model, type, spec. *selection*	Purchasing
2	Power source	Repair	Electrical power loss	Black out	10	Overload	2	Safeguarding system	1	20	Preventive maintenance	General maintenance
3	Office /house electrical wiring	Repair	Electrical power loss	Black out	10	Wiring problem	2	Annual wiring check	1	20	Preventive maintenance	General maintenance
4	Electrical accessories	Replace	Elect. switch broken	Jammed	4	Broken switch	1	As request	1	4	Preventive maintenance	General maintenance

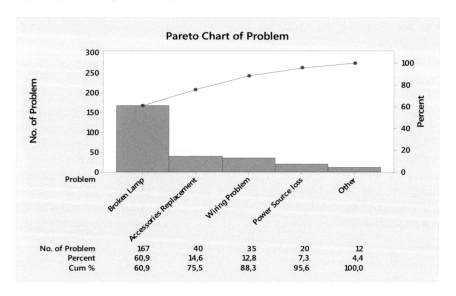

Fig. 2 Pareto chart of maintenance activity with electrical category

dominant factor from the calculation as shown in Table 3. Referring to Table 3, the recommended actions were lamp model, type, and specifications selection.

In this company, more than 10,000 of 36 W incandescent tube light are widely used for housing and building maintenance under the general maintenance division. This was decided as the case study to illustrate the proposed 16 W LED tube light model type and specifications by using LCC calculation. The proposed type is the equal watt comparison to the current tube light referring to the manufacture technical specifications. After the lamp has been used for a certain period of time or at the end of its useful life, the C_{ic} and C_{omc} data are calculated to get the total C_{ic} (TC_{ic}) and C_{omc} (TC_{omc}) of all important sections (Table 4) for calculating the total LCC and prioritizing the order of cost consumption for the cost of one tube lamp, life expectancy, electricity consumption, or failure mode which can be useful for developing purchasing strategies.

Table 4 Calculating total LCC

Alternatives	Count	Fluorescent tube lamp brand A	Count	LED tube lamp brand A	Count
Initial cost	1	40	1	85	1
Operation and maintenance cost	1	534	1	233	1
LCC	1	574	1	322	1
N =	3	N =	3	N =	3

6 Life-Cycle-Cost Analysis

Referring to the results from Table 4 and Fig. 3, the tube lamp was calculated using the formula (LCC $= C_{ic} + C_{omc}$). The following attractive findings were found: (1) The purchasing price of current (fluorescent) tube lamp is 47% cheaper than the proposed (LED) tube lamp but 229% higher LCC than the proposed tube lamp with the same brand. (2) The operational and maintenance costs (C_{ome}) are significantly higher compared to the initial cost items. However, this was always ignored in the past without scientific calculations. The cost–benefit analysis using the life cycle cost of the tube lamp type shows that the proposed type is more beneficial than the current type. This finding can be used as one of the parameters in order to obtain a significant cost-beneficial improvement in the purchasing strategy at the next purchasing decision such as referring to the cost comparison shown in Fig. 3.

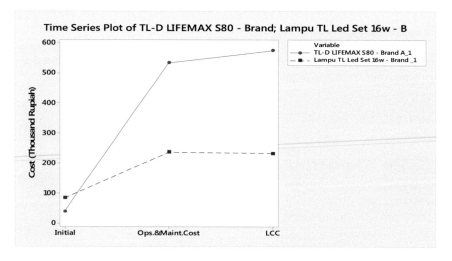

Fig. 3 LCC comparison

Table 5 Life-cycle-cost analysis

year	Count	Fluorescent tube lamp brand A	Count	LED tube lamp brand A	Count
0	1	40.00	1	85.00	1
1	1	351.84	1	223.59	1
2	1	703.67	1	447.19	1
3	1	1055.51	1	670.78	1
4	1	1407.35	1	894.38	1
5	1	1759.19	1	1117.97	1
6	1	2111.02	1	1341.57	1
7	1	2462.86	1	1565.16	1
8	1	2814.70	1	1788.75	1
9	1	3166.54	1	2012.35	1
10	1	3518.37	1	2235.94	1
N =	11	N =	11	N =	11

7 Conclusion

This study found that purchasing a tube lamp using the LCC calculation into account gives a better choice than only considering the acquisition price. This case study uses the FMEA application to prioritize accurate technical activities. LCC is used for measuring costs, initial costs, and operational and maintenance costs (C_{ome}). The calculation results can assist the company in deciding and purchasing better materials based on quality and budget. This will be expanded by purchasing management design to increase competition throughout the supply chain. Estimates of the total cost of tube lamps in the minimum equipment cycle are applied to assess the supplier. LED tube lamps are IDR 251,649 or 43.8% cheaper over their lifetime compared to fluorescent tube lamps. Therefore, a good economist would prefer to buy LED tube lamps for this company for long-term use as it is proven for a better economical beneficial in the long run.

References

1. F. Zachariassen, J. Stentoft Arlbjørn, Exploring a differentiated approach to total cost of ownership. Ind. Manag. Data Syst. **111**(3), 448–469 (2011)
2. M. Góralczyk et al., LCC application in the polish mining industry. Manag. Environ. Qual. **16**(2), 119–129 (2005)
3. D.G. Woodward, Life cycle costing—theory, information acquisition and application. Int. J. Proj. **15**(6), 335–344 (1997)
4. P.C. Teoh, K. Case, Failure modes and effects analysis through knowledge modelling. J. Mater. Process. Technol. **153–154**, 253–260 (2004)
5. M. Murphy et al., A methodology for evaluating construction innovation constraints through project stakeholder competencies and FMEA. Constr. Innov. **11**(4), 416–440 (2011)

6. T. Xuguan, X. Wei, Improving product quality based on QFD and FMEA theory-taking electro-fusion joints as an example, in *International Conference on Computer Science and Information Engineering* (ICCSIE 2018), (CSP)
7. D.S. Chang, K.L. Paul Sun, Applying DEA to enhance assessment capability of FMEA. Int. J. Qual. Reliab. Manag. **26**(6), 629–643 (2009)
8. C. Dong, Failure mode and effects analysis based on fuzzy utility cost estimation. Int. J. Qual. Reliab. Manag. **24**(9), 958–971 (2007)
9. I.B. Utne, Life cycle cost (LCC) as a tool for improving sustainability in the Norwegian fishing fleet. J. Clean. Prod. **17**(3), 335–344 (2009)
10. J. Lutz et al., Life-cycle cost analysis of energy efficiency design options for residential furnaces and boilers. Energy **31**(2–3), 311–329 (2006)
11. T.L. Lamers et al., Application of a modified quality function deployment method for MEMS. Int. Mech. Eng. Congr. Expo. **11**, 159–168 (2007)
12. Y. Akao, G.H. Mazur, The leading edge in QFD: past, present and future. Int. J. Qual. Reliab. Manag. **20**, 20–35 (2003)

Preliminary Study and Analysis on the Electricity Consumption of a Library Building

Norasikin Hussin, Yusli Yaakob, Azli Abd Razak, and Faizal Baharum

Abstract Energy consumption in higher education buildings is huge due to their large number of buildings and population. Within the educational buildings, university campuses have complex functions by providing space for various activities and disciplines. University campuses with library building have different features and changing energy needs compared to other buildings. The electricity consumption in the building comes from the use of air conditioning systems including air handling unit operations, lighting, computers, office equipment and elevator systems. In this research, energy consumption analysis was conducted at Perpustakaan Tun Abdul Razak (PTAR) at Permatang Pauh campus. This project is aims to determine the energy consumption of library buildings and to analyze the connection between the indoor temperature and humidity. The study is based on the electricity consumption by using the Fluke 1750 with data logger. It is found that the energy consumption of chillers system is to be around 46.7%, lighting and sockets follow with 22.5%, AHU with 19.7 and 11% for others. This study shows that the indoor temperature and humidity have the highest influence on the building energy consumption.

Keywords Library building · Energy consumption · Electricity · Air conditioning

N. Hussin · Y. Yaakob (✉)
Faculty of Mechanical Engineering, Universiti Teknologi MARA Cawangan Pulau Pinang, Kampus Permatang Pauh13500 Permatang Pauh, Pulau Pinang, Malaysia
e-mail: yusli662@uitm.edu.my

N. Hussin
e-mail: norasikin245@uitm.edu.my

A. A. Razak
Faculty of Mechanical Engineering, Universiti Teknologi MARA Shah Alam, Shah Alam, 40450 Selangor, Malaysia
e-mail: azlirazak@uitm.edu.my

F. Baharum
School of Building, Housing and Planning, Universiti Sains Malaysia, 11800 Pulau Pinang, Malaysia
e-mail: faizalbaharum@usm.my

1 Introduction

Buildings in Malaysia consume 14.3% of the overall energy and 53% of electrical energy is consumed in residential and commercial sectors [1]. One of the commercial sectors is contributes to the high cost of electricity consumption every year is educational building. Most of the educational buildings use a large electricity to operation and maintenance of electrical and mechanical systems such as air conditioning systems, lighting systems, office equipment and elevator systems. Air conditioning systems provide the people working/living inside buildings with the temperature and humidity and movement of the air should be within certain acceptable comfort ranges [2]. The energy consumption in educational buildings depends on the buildings activities [3]. P. Hammed et al. have pointed out that 90% of people spent most of their time inside buildings and effects to increasing the building energy consumption and thermal environment[1]. A study by Jin Zhou et al.shows that the correlation of energy consumption under cooling system and indoor thermal comfort condition has a strong correlation and significance [4]. The increasing of building cooling load will give a big impact on building energy performance [5]. The electricity consumption in buildings was from air conditioning systems around 50%, while 20% of lighting systems and the rest is plug loads [6]. The operation of the air handling unit (AHU) in an air conditioning system also affects energy consumption [7]. AHU is device used to regulate and circulate air as part of cooling system. Hence, the electricity is still an essential input for the building and a necessity to operate the mechanical and electrical system/equipment. In hot and humid climates, the use of air conditioning systems is considered a major factor in increasing energy consumption in buildings. In hot weather as well, heat gains from the sun, number of occupants, lighting and appliance systems are considered to influence the use of air conditioning systems in buildings. The buildings located in hot and humid climates required high energy use to operate air conditioning system [8]. Bruan et al. [9] investigate the effect of outdoor temperature and humidity on the buildings electricity. It was found that temperature has the highest influence on the building energy consumption. The work of Bahman et al. [10] shows that the indoor relative humidity is strongly correlated with total energy consumption. The authors claimed that the lower indoor humidity, the lower the total energy use.

In this study, the energy performance of the educational building located in the Penang, Permatang Pauh campus, was evaluated. This building is located in hot and humid climate where high energy consumption is used for cooling system. Therefore, this research aims to investigate the pattern of electricity consumption in buildings by using Fluke 1750 with data logger. Indoor temperature and humidity in the building have been measured. As a result of the case study, the relationship between indoor space conditions with energy consumption will be identified.

2 Methodology

In order to investigate the amount of electricity consumption in the library building, the measuring in the main switchboard (MSB) library has been conducted to measure the air conditioning electrical consumption and thermal comfort conditions inside the building. In order to analyze the energy consumption of the building, the real-time (actual) data was collected from October 2018 to January 2019. The Fluke 1750 with data logger has been installed in the MSB and each of its distribution in order to find out the electricity consumption. The device has been used to record the electricity consumption of the building. All the recording data is automatically stored on the internal flash memory. Analysis can be performed by using the Fluke Power Analyze software.

Figure 1 shows the electricity distribution of the building. As shown in Fig. 1, the main source of electricity university building was supplied to the MSB library building and distributed to distribution boxes (DB). The electricity from MSB library was supplied to eleven distribution boxes (DB) of PTAR library and one DB for Balai Kor Suksis building. Each of DB library building provides electricity support for the different functions such as MDB-G, MDB-1, MDB-2, AHU-G, AHU-1, AHU-2, ACMV, Feeder Pillar, Pneumatic Pump and Booster Pump. Furthermore, the main distribution box (MDB) for each floor supplies electricity for lighting system and sockets. The electricity from MSB library building is distributed to air conditioning and mechanical ventilation (ACMV) box to operate the centralized chiller system. The AHU box supplied electricity for air handling unit for each floor. AHU is a device used to regulate and circulate air as part of cooling system. The essential main switchboard (EMSB) was divided into several subsections including elevator system and fire-fighting system.

3 Case Study

The campus building consists of the Laman Perdana, Perpustakaan Tun Abdul Razak (PTAR) library, College Academic Building and College Classroom (BKBA), Pusat Islam, Dewan Besar, Residential Colleges, and the Balai Kor Suksis building. The PTAR library is a standalone and three-story building that was used for this case study (see Fig. 2). The PTAR library building is located in Permatang Pauh campus, at latitude 5.3°N and longitude 100.27°E. The major energy consumers of this building are the air conditioning system, lighting system, elevator system and office equipment. The operation hours of building are from 8.30 am to 9.45 pm (Monday to Friday) and from 8.00 am to 4.45 pm (Saturday). The PTAR library building has closed on Sunday and Malaysia public holiday.

The centralized air conditioning system was operated from 7.30 am. Cooling system has three units of chiller with the capacity of 120 ton refrigeration (tr) for each. The chillers are controlled by the cooling demand of the AHUs. The setting

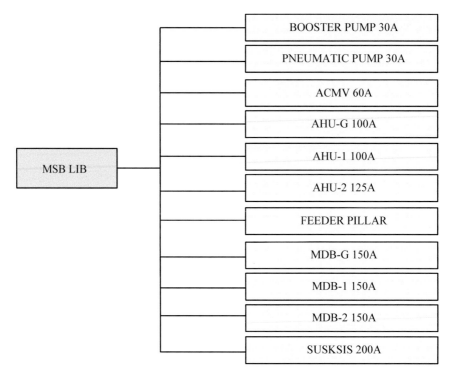

Fig. 1 Electricity distribution of the PTAR library

Fig. 2 PTAR library building

Table 1 Library building system details

System/equipment	Detail category	Operating hours
Cooling system	Air-cooled centralized	7.30 am–9.00 pm (Monday to Friday) 7.30 am–5.00 pm (Saturday)
Lighting	Fluorescent	8.30 am–9.45 pm (Monday to Friday) 8.00 am–4.45 pm (Saturday)
Office equipment	Standard	

of thermostat control was 20 °C. The specification of the system, equipment and operating hours is given in Table 1.

4 Results and Discussion

The objectives of this research is to investigate the electricity consumption in PTAR library building. The measurement of electricity consumption in MSB by using Fluke 1750 with data logger has been carried out from October 2018 to January 2019. The measured data in the MSB library building shows that the electricity consumption in DB Balai Kor Suksis building is less compared to DB library building. More electricity is consumed in the library building to operate the electrical and mechanical systems. The electricity consumption breakdown for the library building is shown in Fig. 3. It can be seen from Fig. 3 that the largest amount of electricity (46.72%) is consumed by the chillers system while lighting and sockets system consumes 22.52%, AHU system consumes 19.72% and others is 11.04% such as elevator system and fire-fighting system. The finding shows that the library building consumes electricity in average of 3855.78 kWh. The centralized chiller system is the largest energy-consuming compared to lighting system and office equipment.

A study by Ying Hang et al. [11] claimed that the air conditioning in the library building is the highest energy consumption in university building. The function of air conditioning system is to maintain the building at indoor temperature and humidity and remove the heat gain. Thus, the design of cooling load is the amount of heat energy to be removed by the air conditioning system [12]. The air conditioning energy consumption of the buildings is significantly related to the type of cooling [13]. The reading of electricity consumption in MSB library building has been measured by Fluke 1750 with data logger, and the analysis has been performed using the Fluke Power Analyze software. The analysis is conducted to ensure the chiller control system is to meet the cooling load demand while minimizing electricity consumption. The results of weekday's energy consumption are shown in Fig. 4. From the figure, we can see that the electricity consumption is inconsistent and start to increase between 7:00 am to 7:30 am. According to Table 1, the cooling system has started functioning at 7:30 am. Because of that, the reading of electricity consumption has increased rapidly from 7:30 am to 8:00 am which is from 17.41 to 121.04 kWh. As

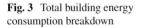

Fig. 3 Total building energy
consumption breakdown

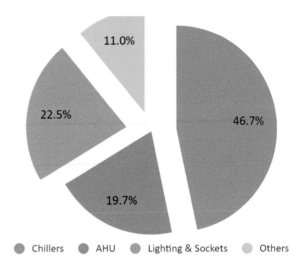

shown in Fig. 4, the profile shows that the reading of electricity consumption is fluctu-
ated when the chiller system is operated. The weekday's consumption of electricity is
rangging between 110 to 122 kWh. Previous studyhave found that fluctuation electricity is commonly influenced by the cooling system [13], thermal comfort [14],
building activities and climate [8]. Average daily electricity consumption and indoor
temperature profile for weekdays are shown in Fig. 5. The measurement of indoor
temperature and humidity has been measured using Barometer with data logger. The
result shows that the indoor temperature fluctuates depending on the time.

Figure 5 shows that more electricity consumption is needed to cool a building when
the chiller system starts to run. For example, the electricity consumption is high at
8.00 am within 121 kWh. At this period, the temperature of the building is 23.9 C.
The electricity consumption decreases when the temperature is decreased. At 10 am,
the electricity consumption is 111 kWh, while the indoor temperature is 23.3 °C.

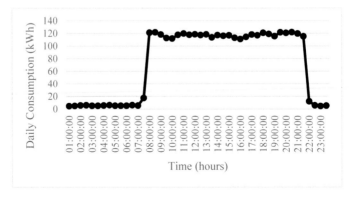

Fig. 4 Average daily electricity consumption profile for weekdays

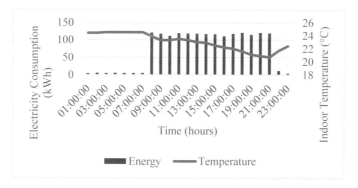

Fig. 5 Average daily electricity consumption and indoor temperature profile for weekdays

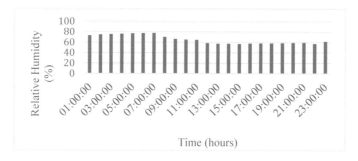

Fig. 6 Average daily humidity profile for weekdays

Thus, the fluctuation of indoor temperature has an effect to electricity consumption. During hottest days, indoor temperature can influence the cooling load. High indoor air temperature and relative humidity affect the occupants comfort and cooling load of the building. Figure 6 shows the daily average humidity profile for weekdays in the building. As shown in Fig. 6, the results show that the range of relative humidity during working days is 56–70%. The building depends on mechanical systems to achieve required levels of comfort. Building occupant behavior and lifestyle consume high energy usage for daily activities and comfort conditions [15]. Indoor temperature and specific humidity are influenced to building design and structures on electricity consumption [16].

5 Conclusion

In this study, we investigated the energy consumption in the library building by using Fluke 1750 with data logger. The finding shows that the energy consumption of chillers system is to be around 46.7%, lighting and sockets follow with 22.5%,

AHU with 19.7% and 11% for others. This study shows that the indoor temperature and humidity have the highest influence on the building energy consumption.

Acknowledgements The authors would like to acknowledge PTAR, Permatang Pauh campus which provided the tools required for conducting this research.

References

1. P. Hameed et al., Building energy for sustainable development in Malaysia : a review. Renew. Sustain. Energy Rev. **75**, 0–1 (2016).https://doi.org/10.1016/j.rser.2016.11.128
2. J.C. Solano, E. Caamaño-Martín, L. Olivieri, D. Almeida-Galárraga, HVAC systems and thermal comfort in buildings climate control: an experimental case study. Energy Rep. **7**, 269–277 (2021). https://doi.org/10.1016/j.egyr.2021.06.045
3. I. El-Darwish, M. Gomaa, Retrofitting strategy for building envelopes to achieve energy efficiency. Alexandria Eng. J. **56**(4), 579–589 (2017). https://doi.org/10.1016/j.aej.2017.05.011
4. J. Zhou, Y. Song, G. Zhang, Correlation analysis of energy consumption and indoor long-term thermal environment of a residential building. Energy Procedia **121**, 182–189 (2017). https://doi.org/10.1016/j.egypro.2017.08.016
5. W.I.W. Nazi, Y.D. Wang, T. Roskilly, Methodologies to reduce cooling load using heat balance analysis: a case study in an office building in a tropical country. Energy Procedia **75**, 1269–1274 (2015). https://doi.org/10.1016/j.egypro.2015.07.185
6. J. Litardo, R. Hidalgo-Leon, G. Soriano, Energy performance and benchmarking for university classrooms in hot and humid climates. Energies **14**(21), 1–17 (2021). https://doi.org/10.3390/en14217013
7. J. Gao, Y. Sun, J. Wen, T.F. Smith, An experimental study of energy consumption and thermal comfort for electric and hydronic reheats. Energy Build. **37**(3), 203–214 (2005). https://doi.org/10.1016/j.enbuild.2004.05.012
8. M.M. Rahman, M.G. Rasul, M.M.K. Khan, Energy conservation measures in an institutional building in sub-tropical climate in Australia. Appl. Energy **87**(10), 2994–3004 (2010). https://doi.org/10.1016/j.apenergy.2010.04.005
9. M.R. Braun, H. Altan, S.B.M. Beck, Using regression analysis to predict the future energy consumption of a supermarket in the UK. Appl. Energy **130**, 305–313 (2014). https://doi.org/10.1016/j.apenergy.2014.05.062
10. A. Bahman, L. Rosario, M.M. Rahman, Analysis of energy savings in a supermarket refrigeration/HVAC system. Appl. Energy **98**, 11–21 (2012). https://doi.org/10.1016/j.apenergy.2012.02.043
11. Y. Han, X. Zhou, R. Luo, Analysis on campus energy consumption and energy saving measures in cold region of China. Procedia Eng. **121**, 801–808 (2015). https://doi.org/10.1016/j.proeng.2015.09.033
12. V. Kumar, J. Liman, S.A. Alkaff, Science direct science direct comparative study of passive methods for reducing cooling load assessing the feasibility of using the heat demand-outdoor temperature function for a long-term district heat demand forecast rise Pina is Corre. Energy Procedia **142**, 2689–2697 (2017). https://doi.org/10.1016/j.egypro.2017.12.212
13. L. Yuan, Y. Ruan, G. Yang, F. Feng, Z. Li, Analysis of factors influencing the energy consumption of government office buildings in Qingdao. Energy Procedia **104**, 263–268 (2016). https://doi.org/10.1016/j.egypro.2016.12.045
14. M. Indraganti, D. Boussaa, Comfort temperature and occupant adaptive behavior in offices in Qatar during summer. Energy Build. **150**, 23–36 (2017). https://doi.org/10.1016/j.enbuild.2017.05.063

15. N. Jamaludin, N.I. Mohammed, M.F. Khamidi, S.N.A. Wahab, Thermal comfort of residential building in Malaysia at different micro-climates. Procedia Soc. Behav. Sci. **170**, 613–623 (2015). https://doi.org/10.1016/j.sbspro.2015.01.063

16. H. Guan, S. Beecham, H. Xu, G. Ingleton, Incorporating residual temperature and specific humidity in predicting weather-dependent warm-season electricity consumption. Environ. Res. Lett. **12**(2) (2017). https://doi.org/10.1088/1748-9326/aa57a9

A Compact Rectangular Slotted Microstrip Bandpass Filter

S. M. Norzeli, Erol Terović, Şehabeddin Taha İmeci, N. Ramli, R. Shafie, and S. M. Sharun

Abstract This paper presents a compact rectangular slotted microstrip bandpass filter in a 3D planar electromagnetic simulation software called Sonnet Suites. The microstrip filter design is symmetric along the inverse diagonal and consists of a dual-mode microstrip square ring resonator with an alternative perturbation element. This design has a compact rectangular structure of 1×1 cm. The substrate used is FR-4 with a dielectric constant of $\varepsilon r = 4.4$ and a thickness of 1.6 mm. The filter has a center frequency of 4.33 GHz and a bandwidth of 1.9 GHz, with S_{12} is nearly zero while S_{11} is -23 dB. This filter can be implemented between 3 and 5 GHz operating frequencies in wireless.

Keywords Microstrip bandpass filter · FR-4 · Microstrip · Rectangular · Symmetry

S. M. Norzeli (✉) · S. M. Sharun
Faculty of Innovative Design and Technology, Universiti Sultan Zainal Abidin, 21300
Terengganu, Malaysia
e-mail: syamiminorzeli@unisza.edu.my

S. M. Sharun
e-mail: sitimaryam@unisza.edu.my

E. Terović · Ş. T. İmeci
Faculty of Engineering and Natural Sciences, International University of Sarajevo, Hrasnicka
Cesta 15, Ilidža, 71210 Sarajevo, Bosnia and Herzegovina
e-mail: simeci@ius.edu.ba

N. Ramli
Centre for Advanced Electrical and Electronic System (CAEES), Faculty of Engineering, Built
Environment and Information Technology, SEGi University, Petaling Jaya, Selangor, Malaysia
e-mail: azlinaramli@segi.edu.my

R. Shafie
School of Electrical Engineering, College of Engineering, Universiti Teknologi MARA, 23000
Dungun, Terengganu, Malaysia
e-mail: rosma5455@uitm.edu.my

M. H. Abu Bakar et al. (eds.), *IT Solutions for Sustainable Living*,
SpringerBriefs in Applied Sciences and Technology,
https://doi.org/10.1007/978-3-031-51859-1_8

1   Introduction

Most communication systems require a filter used in analog signal processing. Microstrip filters commonly use devices operating in the 800 MHz to 30 GHz frequency range [1–3]. There are two types of filters: parallel-coupled line filters and interdigital filters [4]. A bandpass filter is a passive component that can select signals that belong to a specific bandwidth with a particular center frequency while simultaneously rejecting other signals that belong to different frequencies [5]. Microstrip bandpass filter design has been studied for over 50 years [6]. Modern microwave communication systems need microstrip bandpass filters that have improved return loss responses, high rejection, reduced size and good insertion loss [7].

A design method using the Sonnet Suites software [8] that employs a rigorous Method-of-Moments (MoM) Electromagnetic (EM) analysis, based on Maxwell's equations is presented in this article. In this paper, a new topology for a bandpass filter is proposed. The filter design is simulated and iteratively improved until a usable bandpass characteristic is obtained by observing the S parameters of the filter. Once a good result has been obtained, the filter was fabricated, and a good agreeance was obtained between the simulation results and the filter manufactured using the FR-4 substrate. As mobile communication systems become more widespread, bandpass filter development has emphasized compactness and high performance [9].

In [10], designs using traditional bandpass filter topologies like the parallel-coupled microstrip bandpass filter, where strips are arranged beside each other and are coupled. Different filter topologies have been proposed, such as a slot-coupled diamond shape [11], a ring resonator with a stub [12] or a filter using a meander loop resonator [13]. This project uses a modified version of the filters with a single ring with one or two tuning stubs [14] and a ring UWB bandpass filter [15]. Additionally, this filter uses coupled feed lines and perturbation elements [16, 17] introduced a planar dual-mode wideband bandpass filter design. A ring resonator with no perturbation element was employed in the filter. It also included feed lines that were directly connected [18].

2   Filter Design

In this paper, the filter is designed symmetrically along the inverse diagonal. It is a modified dual-mode microstrip square ring resonator with an alternative perturbation element. Figure 1 shows the topology of the design, and Fig. 2 represents the 3D view of the filter topology. The filter response can be seen in Fig. 3, with the red graph representing the reflection coefficient S_{11} and the blue graph representing the forward transmission coefficient S_{21}.

Fig. 1 Top view of the filter
with dimensions

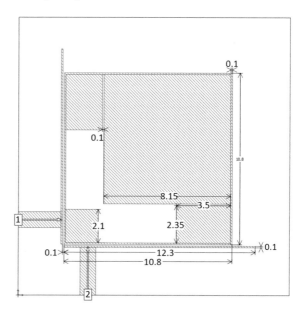

Fig. 2 The 3D view of filter
topology

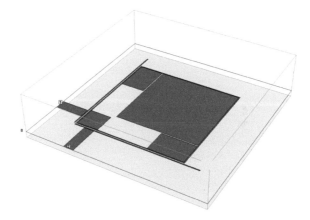

3 Parametric Study

The first step of the parametric study was done by focusing on the two small rectangles
that started with dimensions of 2.1×0.5 mm. The size was increased in increments of
0.5 mm. Since the filter is symmetrical and inverse diagonal, the dimensions shall be
given for the top left rectangle and be the same for the bottom right, except width and
height are switched. Tables 1, 2 and 3 contain data showing a reflection coefficient
of -10 dB to infinity. This topology was then modified in the parametric study.

Table 1 shows the results of the first parametric study. The length of the two small
rectangular structures has been changed from 3.5 to 6 mm. The goal was to maximize

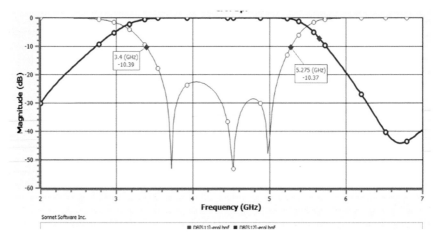

Fig. 3 Filter response graph

Table 1 Optimization of the two small rectangular structures

Dimensions (mm)		Magnitude (dB)		Bandwidth (GHz)
Width	Length	S_{11}	S_{21}	
2.1	3.5	−10.380	−0.488	1.615
2.1	4.0	−11.730	−0.312	1.572
2.1	4.5	−10.55	−0.390	1.558
2.1	5	−11.65	−0.293	1.436
2.1	5.5	−11.390	−0.392	1.314
2.1	6.0	−10.22	−0.424	1.238

Table 2 Optimization of a rectangular gap between the two small rectangles and the large square structure	Gap (mm)	Magnitude (dB)		Bandwidth (GHz)
		S_{11}	S_{21}	
	0.1	−13.410	−0.195	1.747
	0.2	−12.140	−0.228	1.707
	0.3	−10.710	−0.488	1.633

Table 3 Dielectric thickness optimization

Dielectric thickness (mm)	Magnitude (dB)		Bandwidth (GHz)
	S_{11}	S_{21}	
1.5	−21.970	−0.032	1.873
1.53	−22.400	−0.031	1.884
1.55	−22.850	−0.023	1.901
1.57	−22.900	−0.027	1.905
1.6	−23.240	−0.018	1.914

Fig. 4 Current density simulation

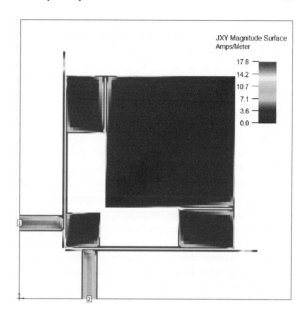

the bandwidth for the application. The rectangle dimensions of 2.1 × 3.5 mm have been chosen as the best result with the highest bandwidth.

Table 2 shows the optimization of a rectangular gap between the two small rectangles and the large square structure with width and length fixed at 2.1 and 3.5 mm, respectively. The gap increased by 0.1 mm and incremented by 0.1 mm. The best parameter is a 0.1 mm gap with the highest bandwidth of 1.747 GHz. Only three results are shown because all the results above 0.3 mm yielded a poor reflection coefficient, rendering the filter unusable.

The dielectric thickness of the substrate was optimized between 1.5 and 1.6 mm, as shown in Table 3. According to the results, the bandwidth increases as the dielectric thickness increases. Therefore, dielectric thickness with 1.6 mm has been chosen due to the highest bandwidth.

Figure 4 depicts the current density simulation. The center frequency was chosen for the current density simulation. The color bar with the A/m equivalent is given in the upper right corner of Fig. 4.

4 Conclusion

In conclusion, the paper proposed a novel design for a compact rectangular slotted microstrip bandpass filter. The bandpass filter shows good performance in its operation area, and alongside this, it remains compact and simple. The bandwidth of this filter was 1.914 GHz, with an S_{11} of −23.24 dB and an S_{21} of −0.018 dB. Different parametric studies regarding the length of the two small rectangular structures, the rectangular gap and the dielectric thickness were optimized in this study. These parametric studies show that the filter can be modified to have different characteristics based on specific requirements needed.

References

1. F.H. Wee et al., Electromagnetic wave detection based on multiband antenna design. ARPN J. Eng. Appl. Sci. **11**(8), 4429–4933 (2016)
2. Y.B. Seok et al., Effect of biomass waste filler on the dielectric properties of polymer composites. Environ. Asia **9**(2), 134–139 (2016)
3. S.N.A. Jabal et al., Carbon composition surface porosities and dielectric properties of coconut shell powder and coconut shell activated carbon composites. ARPN J. Eng. Appl. Sci. **11**(6), 3832–3837 (2016)
4. C.J. Kikkert, in *A Design Technique for Microstrip Filters, 2nd International Conference on Signal Processing and Communication Systems, Microwave Symposium Digest* (2008)
5. M. Alaydrus, Designing microstrip bandpass filter at 3.2 GHz. Int. J. Electr. Eng. Inform. **2**(2), 71–83 (2010)
6. S.B. Cohn, Parallel-coupled transmission-line-resonator filters. IRE Trans. Microw. Theory Tech. **6**(2), 223–231 (1958)
7. Q. Yang, X.Z. Xiong, Y. Wu, L.P. Wang, H.Q. Xiao, Design of microstrip tapped-hairpin dual-band pass filter for Ku-band application. Int. Conf. Microwave Millimeter Wave Technol. **2010**, 772–774 (2010)
8. SONNET Suites, High Frequency Electromagnetic Software, User's Guide, release 16, Sonnet Software, Inc (2018)
9. C.-F. Chen, T.-Y. Huang, R.B. Wu, Design of microstrip bandpass filters with multiorder spurious-mode suppression. IEEE Microw. Theory Tech. **53**(12), 3788–3793 (2005)
10. G.A. Hussain, Design of parallel coupled microstrip band-pass filter. Int. J. Compul. Technol. **15**(5), 6768–6775 (2016)
11. H. Chang, W. Sheng, J. Cui, Design of slot-coupled diamond-shape microstrip wideband bandpass filter, in *IEEE MTT-S International Wireless Symposium (IWS)* (2018), pp 1–3
12. H. Ishida, K. Araki, A design of tunable UWB filters, in *International Workshop on Ultra Wideband SYSTEMS Joint, Conference International Journal Ultra Widebd Communication System* (2004), pp. 424–428
13. J.S. Hong, M.J. Lancaster, Microstrip bandpass filter using degenerate modes of a novel meander loop resonator. IEEE Micro. Wirel. Compon. Lett. **5**(11), 371–372 (1995)
14. J.-S. Hong, M.J. Lancaster, Design of highly selective microstrip bandpass filters with a single pair of attenuation poles at finite frequencies. IEEE Trans. Microw. Theory Tech. **48**(7), 1098–1107 (2000)
15. M.H. Weng, S.K. Liu, Y.T. Liu, C.Y. Hung, A low-loss ring UWB bandpass filter. Microw. Opt. Technol. Lett. **50**(9), 2317–2319 (2008)
16. K. Chang, F. Hsu, J. Berenz, K. Nakano, Find optimum substrate thickness for millimeter-wave GaAs MMICs. Microwave Radio Freq. **27**, 123–128 (1984)
17. K. Chang, *Microwave ring circuits and antennas* (Wiley, New York, 1996)
18. M.H. Weng, S.K. Liu, Y.T. Liu, C.Y. Hung, A novel ring-wide bandpass filter by using orthogonal feed structure. Microw. Opt. Technol. Lett. **50**(8), 2025–2027 (2008)

Improvement of Selective Mapping Performance Using Riemann Matrix for PAPR Reduction

A. A. A. Wahab, N. Qamarina, S. S. N. Alhady, W. A. F. W. Othman, and H. Husin

Abstract Wireless communication is one of the most important technologies and something that we use in our everyday life. Most of the technologies that existed use OFDM as a medium to provide their services due to the advantages given such as high data rate, high spectral efficiency, resistance to distortion and many more. However, OFDM suffers from high PAPR that could lead the system to work in nonlinear region and introduce distortion to the system. Hence, it is crucial for OFDM to find a solution to this problem. In this paper, we proposed to SLM as PAPR reduction technique. We also used Riemann Matrix to optimize the phase sequence in SLM.

Keywords OFDM · PAPR · SLM · Riemann matrix

1 Introduction

These days, there are high demand for high-quality wireless services such as higher data rates, larger network capacity, higher energy efficiency and lower latency [1]. To improve the performance of wireless communication system, a multicarrier modulation technique called OFDM is used. This technique uses multiple subcarriers that are orthogonal to each other to transmit data in parallel form. Orthogonal subcarriers are required in large value to support high values of bits/s [2]. OFDM offers a lot of benefits to the system such as robustness to multipath fading, resistance to impulse noise, high spectral efficiency, low sensitivity to time synchronization error, bandwidth flexibility, ability to coexist sharing band with other system, compatible with MIMO and low complexity equalizer. Multipath fading occurs from multiple paths of received signals [3]. Every subcarrier in OFDM is modulated by low-rate data by

A. A. A. Wahab (✉) · N. Qamarina · S. S. N. Alhady · W. A. F. W. Othman · H. Husin
School of Electrical and Electronic Engineering, Universiti Sains Malaysia, 14300 Nibong Tebal, Pulau Pinang, Malaysia
e-mail: aeizaal@usm.my

W. A. F. W. Othman
e-mail: wafw_othman@usm.my

a modulation technique such as QAM or QPSK. These signals will then be converted into time domain to maintain subcarrier's orthogonality by inverse fast Fourier transform (IFFT). To avoid inter-symbol interferences due to multipath fading, cyclic prefix (CP) is added at the beginning of every symbol [4]. A lot of wireless standards have been used OFDM such as LTE, WiMAX, DAB and DVB [1, 5].

OFDM also has some limitations, but the major drawback is high value of peak to average power ratio (PAPR). In advance technologies such as LTE and 5G, it is required for OFDM to work flawlessly. Mm-wave frequency was known to work poorly and introduce high-level distortion in saturated region. When high PAPR signal passing through a small operating region, the signal will be clipped and introduce distortion [2]. So, the issue of high value of PAPR is a big concern since it can lead power amplifiers to work in nonlinear region with low power back off due to large magnitude and cause in-band and out-of-band distortions [5]. Bit error ratio (BER) will face degradation, and to deliver the high power, gain from power amplifier is expensive and not cost effective for the system [4]. PAPR happens when the modulated subcarriers of OFDM subjected to IFFT, a subset of this symbols were linearly added together and produce signal with large peak [1, 6]. Average power of the signal can be reduced to avoid power amplifier working in saturation region but it will reduce signal to noise ratio (SNR) and degrade BER so, it is better to reduce the peak power instead [6].

Over the years, there are a lot of proposed reduction techniques, but all techniques come at their own costs such as loss of data rate, growth in signal power, degradation of BER, increase in complexity of system design, increase of bandwidth, cost inefficient and insertion of another distortion. Performance of reduction techniques is evaluated by these criteria: ability to reduce PAPR, minimum normal power, bandwidth extension, BER performance, signal distortion level, spectral efficiency and computational complexity [5]. Reduction techniques can be divided into three categories: signal scrambling, signal distortion and hybrid. Some of the techniques are clipping and filtering, coding, partial transmit sequence (PTS), selective mapping (SLM), interleaving, peak insertion (PI), tone injection (TI), tone reservation (TR), companding and discrete wavelet transform (DWT) [1, 4].

Among all these reduction techniques, SLM is known to be distortionless and able to effectively reduce PAPR value. This technique also provides no power increase and data loss [5]. In SLM, a set of data block sequence is generated from input data and multiplied to a different set of phase sequences and data block sequence with the lowest PAPR value will be transmitted. The elements of phase sequence corresponding to the chosen data block sequence need to be transmitted to as side information (SI). The phase sequence is conjugated and multiplied with the received data block sequence to retrieve the original information. From the previous studies, it has been found that transmission of SI increases design complexity with a number of subcarriers and results in power efficiency reduction. To ensure the reliability of SI, it is channel coded resulting in spectral efficiency reduction. Computational complexity also increases with extra IFFT and PAPR calculation. There are many proposed method to combat these issues [1, 4].

Proposed a system where alternative signal sequences are generated by linear combination of additive mapping sequences and the sequence with lowest PAPR value will be chosen for transmission [7]. The proposed method manages to improve PAPR performance, and it was discovered that PAPR performance improves only until a certain number of phase sequences and stops achieving linear growth. A new method called SLM-NFS was introduced in [8] where a new phase sequence was employed based on normalized row symmetric Toeplitz matrix. Simulation result shows improvement in PAPR performance compared to other schemes. Suggested a method inspired by PRS-SLM to improve phase sequences generation [9]. This new method has better performance than PRS-SLM and better systematic design architecture than conventional SLM. Idea is to use PSK modulated signal sequence, orthogonal phase sequences sets and deformation parameter sets to generate more OFDM signals with less inverse fast Fourier transform (IFFT) processors [10]. The simulations show that the proposed method manages to improve PAPR performance but not much improvement in computational complexity. In this paper, we proposed to use Riemann Matrix (RM) to optimize the phase sequences of SLM.

2 System Model

2.1 Orthogonal Frequency Division Multiplexing (OFDM)

Input data are modulated by any modulation scheme and combined by IFFT at transmitter. Input data block with length N can be represented as follows [6]:

$$X = (X_0, X_1, X_2, \ldots, X_{n-1})^T.$$

Discrete OFDM signal in time domain can be represented as [1, 4]:

$$x(t) = \frac{1}{\sqrt{N}} \sum_{n=0}^{N-1} x_n e^{j2\pi \Delta f_n t}; 0 \le t \le NT,$$

where x_n is modulated OFDM signal, N is number of subcarriers, Δf_n is subcarrier spacing and NT is data block period.

2.2 Peak to Average Power Ratio (PAPR)

PAPR can be calculated in discrete or continuous time and represents the relationship between peak power of OFDM signal and the average power [2, 6]. N signals with the same phase added together producing a peak that is N times the average signal. PAPR of signal $x(t)$ can be mathematically defined as [1, 4, 6]:

$$PAPRdB = 10 \log 10 \left(\frac{\max(x(t))^2}{\text{avg}(x(t))^2} \right).$$

The performance of PAPR reduction techniques is indicated by complementary cumulative distribution function (CCDF). CCDF is the complementary for CDF and denotes the probability of PAPR exceeding threshold value [2, 4].

$$CCDF = \Pr(PAPR > PAPR_0) = \left(1 - \left(1 - e^{-PAPR_0} \right)^N \right),$$

where $PAPR_0$ is the threshold value.

2.3   Selective Mapping (SLM)

SLM is part of multiple signaling and probabilistic category [8]. In SLM, data block X is multiplied by phase sequences B. The phase sequences can be set to $\{\pm 1, \pm j\}$ [6]. The number of phase sequences denoted by U and the length of phase sequences are same as the length of X [2].

$$X^u = X.B^u,$$

where X^u represents U different data block the represent the same information as X. X^u with lowest PAPR value will be chosen for transmission and the signal can be represented as [6]:

$$x = \text{argmin}_{0 \leq u \leq U-1} \left(PAPR \left(X^u \right) \right).$$

Using the formula $log_2 U = K$, K bits are required to transmit phase sequences as SI [2]. The OFDM signals can be represented as [5, 8]:

$$x(t) = \frac{1}{\sqrt{N}} \int_{n=0}^{N-1} X_n.B^u e^{j2\pi \Delta f_n t}; 0 \leq t \leq NT.$$

2.4   Proposed Method

RM is generated by removing the first row and column of matrix A as follows [11]:

$$A_{i,j} = i - 1 \text{ if } i \text{ divides}$$

$$A_{i,j} = -1 \text{ otherwise.}$$

Assume RM with size MXM, the elements in R1 would be $\left(\frac{1}{M}\right) * RM$ but direct multiplication of real values in RM with OFDM symbols will cause BER degradation as both peak and average value of OFDM symbols. The elements kth row is either k or -1 while the diagonal elements are from 1 to P where P is the size of RM. When the phase sequence chosen is $2 \leq k \leq U$, the amplitude of OFDM symbols also changes together with the phase; therefore, the average power will not be the same as original OFDM signal [12]. RM uses binary sequences instead of real values, so phase sequences produce phase shift of 0 or π radians. This procedure can be explained by the following equation [11]:

$$R^k(y) = sgn\{R1_{k,y}\},$$

where k is number or rows and y is number of columns while sgn is the signum function. Figure 1 shows the flowchart of the proposed method.

At receiver, to retrieve the original information, we only need to know the row number chosen as phase sequence by using additional bits. This can save the time from transmitting the entire phase sequences [11]. Table 1 shows parameters used by the simulation.

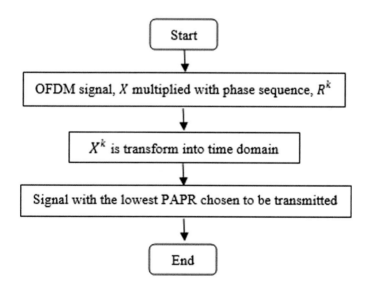

Fig. 1 Flowchart of proposed method

Table 1 Parameters used for simulation

Simulation parameters	Value
Number of subcarriers, N Number of subblocks, M	64, 128, 256, 512 4, 8, 16
Number of phase sequences, U	$\{\pm 1, \pm j\}$
Modulation type	16-QAM

Table 2 PAPR value for different number of subcarriers and phase sequences

Number of subcarrier	Original OFDM	SLM-RM $M = 4$	SLM-RM $M = 8$	SLM-RM $M = 16$
64	16.973	12.809	15.573	15.179
128	20.661	14.906	15.054	15.522
256	21.972	14.866	19.647	18.477
512	21.776	16.155	18.976	20.391

3 Results

Table 2 shows the PAPR performance for different number of subcarriers and subblocks of original OFDM signal and SLM-RM signal. SLM-RM has better performance compared to original OFDM signal. The values of PAPR increase as the number of subcarriers and subblocks increases. Figure 2 shows CCDF performance of original OFDM signal and SLM-RM signal. SLM-RM has the improvement of 1dB compared to OFDM signal. As the number of subcarriers increases, the processing time gets longer due to computational burden. Higher number of subblocks increased computational complexity explaining why PAPR values increase linearly with a number of subblocks.

As can be seen from Fig. 3, there is no BER degradation by using SLM-RM since SM is a distortion less technique. In fact, BER performance gets better by using SLM-RM. The number of subcarriers and subblocks used in Figs. 2 and 3 is 256 and 16, respectively.

4 Conclusions

In this paper, we attempt to use Riemann sequences in SLM to reduce PAPR value in OFDM system. We evaluate the simulation with different number of subcarriers and subblocks. It has been proven in this paper that the method proposed manages to provide a better PAPR and BER performance compared to original OFDM signal. PAPR performances linearly change with different number of subcarriers and subblocks.

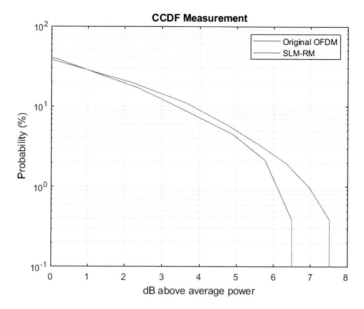

Fig. 2 CCDF performances of PAPR

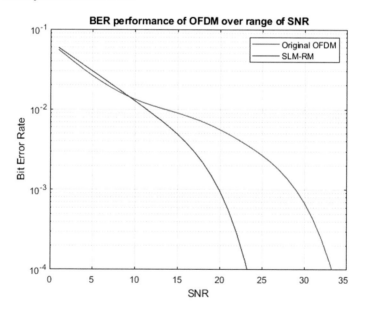

Fig. 3 BER performance

Acknowledgements The authors would like to thank the referees and editors for providing very helpful comments and suggestions. This project was supported by Research University Grant, Universiti Sains Malaysia (203/PELECT/6071476).

References

1. A.M. Rateb, M. Labana, An optimal low complexity PAPR reduction technique for next generation OFDM systems. IEEE Access 16406–16420 (2019)
2. A.B. Kotade, A.B. Nandgaonkar, S.L. Nalbalwar, Peak-to-average power ratio reduction techniques in OFDM: a review and challenges. Int. Conferen. Adv. Commun. Comput. Technol. (ICACCT) **2018**, 319–324 (2018)
3. K. Mahender, K.S. Ramesh, T. Kumar, An efficient of DM system with reduced paper for combating multipath fading. J. Adv. Res. Dynam. Control Syst. (JARDCS) 1939–1948 (2017)
4. S. Sarowa et al., Evolution of PAPR reduction techniques: a wavelet based OFDM approach. Wirel. Pers. Commun. 1565–1588 (2020)
5. R.A. Kumari, M. Chawla, Review of PAPR reduction techniques for 5 G system. Int. J. Electron. Commun. Eng. (AEU) 35–44 (2017)
6. P.P. Ann, R. Jose, Comparison of PAPR reduction techniques in OFDM systems. Int. Conferen. Commun. Electron. Syst. (ICCES) **2016**, 1–5 (2016)
7. J. Karthika, T. Ganesan, R. Mylsamy, PAPR reduction of MIMO-OFDM system with reduced computational complexity SLM scheme. Mater Today Proceed. **37**, 2563–2566 (2020)
8. B. Lekouaghet, Y. Himeur, A. Boukabou, Improved SLM technique with a new phase factor for PAPR reduction over OFDM signals, in *2020 1st International Conference on Communications, Control Systems and Signal Processing (CCSSP)* (2020), pp. 8–12
9. S.Q. You, C.Y. Yang, H.Y. Liang, A modified phase generation mechanism for PAPR reduction in SLM-OFDM systems, in *2020 IEEE International Conference on Consumer Electronics—Taiwan (ICCE-Taiwan)* (2020)
10. Z. Zhou, L. Wang, C. Hu, Improved SLM scheme for reducing the PAPR of QAM OFDM signals, in *2019 IEEE 2nd International Conference on Electronics Technology (ICET)* (2019)
11. S. Mishra, A. Agarwal Peak to average power ratio reduction in sub-carrier index modulated OFDM using selective mapping, in *2017 International Conference on Computer, Communications and Electronics (Comptelix)* (2017)
12. A.Z. Sagar Harne, PAPR reduction in OFDM system using phase sequence of Riemann matrix. Int. J. Adv. Res. Sci. Eng. Technol. 56–60 (2016)

Dual-Band CPW Pentagonal Sierpinski Gasket Fractal Patch Antenna for 3.5 GHz WiMAX and 5.8 GHz WLAN Applications

A. H. A. Rashid, B. H. Ahmad, M. Z. A. Abd Aziz, and N. Hassan

Abstract In this work a dual-band CPW pentagonal Sierpinski gasket fractal antenna is design for WiMAX 3.5 GHz and 5.8 GHz applications. The work has started by Antenna *A* as initiator stage while the improvement stage of Antenna *B* as the first-iteration stage of the proposed antenna is done. The Antenna *A* is designed by a pentagonal-shaped patch, while the Antenna *B* is consisted of the pentagonal patch antenna with the first-iteration slot of the pentagonal shaped. As the investigation done, the Antenna *B* has a sensitivity of—31.70 dB at 3.5 GHz and—22.00 dB at 5.8 GHz, which is acceptable for WiMAX and WLAN operation applications. In this case, this Sierpinski gasket technique that is used in the work had improved the performance of the antenna, focusing on the return loss and effect to shift the resonant frequency.

Keywords Pentagonal antenna · Sierpinski gasket · Fractal geometry · WiMAX · WLAN

A. H. A. Rashid (✉) · B. H. Ahmad · M. Z. A. Abd Aziz · N. Hassan
Faculty of Technology and Engineering Electronic and Computer (FTKEK), Universiti Teknikal Malaysia Melaka (UTeM), 76100 Durian Tunggal, Melaka, Malaysia
e-mail: amier.utem@gmail.com

B. H. Ahmad
e-mail: badrulhisham@utem.edu.my

M. Z. A. Abd Aziz
e-mail: mohamadzoinol@utem.edu.my

N. Hassan
e-mail: nornikman84@yahoo.com

1 Introduction

At the moment, current communication systems require the use of antennas with a wide bandwidth, enhanced efficiency, and robustness during transmission and reception, which is why implementing them using conventional designs has rendered them obsolete in some circumstances. The field of wireless communications necessitates the enhancement of bandwidth without jeopardizing the antenna design's actual functionality.

Microstrip patch antennas were invented in the 1950s [1] but did not achieve widespread adoption until the 1970s, owing to the rapid advancement of integration technology with active devices, which enables the realization of extremely tiny designs. Despite their low profile, small size, low manufacturing costs, and ease of circuit integration, these antennas often have low fractional bandwidths that are incompatible with modern wireless communication systems. Various solutions based on fractal geometries have been lately developed to reduce the size of the antenna or to improve its properties [2, 3].

Fractal geometry has a wide range of applications in the sciences, such as biology, geology, and engineering. Fractal geometries have been employed in the design of antennas, frequency selective surfaces, image processing, and biomedical signal processing in engineering. Mandelbrot in 1977 had been introduced, developing and pioneering fractal geometry, as well as its dimensions in his research work [4]. The term in Latin for 'fractured' is also associated with 'fragmentation' and 'uneven fragments'. Fractals, according to Mandelbrot, are symmetries, which are invariances whether expanded or contracted [5]. Fractal geometry was born out of the study of natural patterns.

It is a complex fractured geometry broken into parts, each of which is a cut-size reproduction of the total design. The self-similarity property of fractal geometry design is a repeating orientation of self-similar objects in each iteration step [6]. Small size, improved input impedance, wideband/multi-band support, consistent performance over a wide frequency range, and the inclusion of inductance and capacitance without the use of additional components are all benefits of fractal geometry structures [7]. Electrically huge features are the result of the space-filling attribute. The ability to use an iteration function system with comparable shapes is made possible by the self-similar property. When an iteration function system has the self-scaling property, it can use similar shapes at various scales. These characteristics allow them to be compactly packed, allowing them to be displayed in small spaces. The Sierpinski gasket [8], Von Koch snowflakes [9], Hilbert curve [10], and Minkowski curve [11] are all examples of fractal geometry in design (Fig. 1).

Numerous improvement parameters have been developed that include fractal geometry into the antenna design. In his study, [12] introduce a new hexagonal-shaped fractal ultra-wideband CPW-fed antenna capable of rejecting the frequency band proposed for IEEE 802.11a and WLAN-HIPERLAN/2 from 5.1 to 5.9 GHz using a U-slot filter. The use of hexagonal slots and fractal geometry at the edges enables the electrical trajectory to be extended and the UWB characteristics to be improved over the frequency range of 3.1–10.6 GHz.

Gorai [13] created a compact UWB planar quasi-self-complementary Von Koch fractal boundary over the hexagonal-shaped radiator in his research to increase the antenna's input impedance matching at 2.4 and 5.2 GHz. Both the electric and magnetic counterparts contain hexagonal-shaped boundaries embedded with the first-iteration Von Koch fractal boundary to aid in impedance matching at higher frequencies.

Radonic [14] demonstrated a Hilbert fractal dipole antenna operating at 2.4 GHz in his work by modifying the line-to-spacing ratio of the Hilbert curves using Low-Temperature Cofired Ceramic (LTCC) technology. This antenna's impedance matching to the input impedance and reflection characteristic have been done.

A noteworthy fractal geometry is the Sierpinski gasket, discovered in 1916 by Waclaw Sierpinski [15]. An initiator, shown in Table 1 as a Euclidean triangle in Step 0, serves as the starting point for the design process.

Sierpinski gasket-inspired fractal multi-band antennas with five distinct heights and a factor of two between scales were pioneered by Puente in 1996 [16]. In this paper, the proposed CPW fractal pentagonal patch antenna for WiMAX and WLAN applications is designed to obtain dual-band frequencies' range of 3.5 GHz and 5.8 GHz. A pentagonal slot is located at the pentagonal patch antenna to improve the performance.

Table 1 Sierpinski iteration development stage, (a) triangular, (b) square

Shape	Initiator, Step 0	First iteration	Second iteration
Euclidean triangle			
Square			

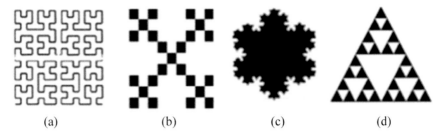

Fig. 1 Example of the fractal geometry design, (**a**) Hilbert curve [10], (**b**) Minkowski [11], (**c**) Koch snowflakes [9], (**d**) Sierpinski gasket [8]

2   Antenna Design

This antenna was built on a FR4 substrate with a thickness of 1.6 mm and a dielectric constant of 4.3 with the copper thickness of 0.035 mm. The initial stage of designing a microstrip antenna began with the antenna's basic construction of pentagonal-shaped patch.

The antenna is consisted of two stages of development, which consist of Antenna *A* and Antenna *B*. The initial stage of designing a microstrip antenna began with the antenna's basic construction of pentagonal-shaped patch. The enhancement design process had been done with the implemented of the pentagonal slot structure at the patch of the antenna. Figure 2 shows the CPW Pentagonal Sierpinski Gasket Fractal Patch Antenna (simulation and fabricated version).

3   Results

This section describes the proposed antenna parameter, which consists of the gain in GHz (dB vs. frequency), return loss in GHz (dB versus frequency), and the bandwidth and radiation pattern of the antenna. For resonance frequency, this antenna must transmit and receive at least 90% of the signal. Figure 3 shows the return loss of Antenna *A* and Antenna *B*.

For Antenna *A*, it shows that the antenna is operated at two different locations of the frequencies at 3.52 GHz and 5.97 GHz with return loss of – 20.71 dB and – 25.17 dB, respectively. The improvement of return loss of the Antenna *B* by introduced a pentagonal slot of iteration 1 is effect to shift resonant frequency of the antenna performance. In this case, the Antenna B had been operating at two different location points of 3.32 GHz and 5.77 GHz with – 21.61 dB and – 49.98 dB, respectively.

Next, it goes to the comparison between simulation and measurement result performances of Antenna *B*. At this case, the measurement had smaller differential of shifting point of the resonant frequency compared with the simulation results. For measurement result, it has operated at 3.55 GHz as the first resonant frequency while

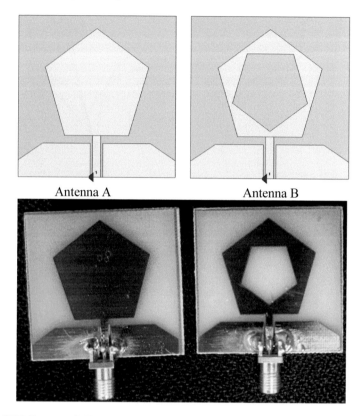

Fig. 2 CPW Pentagonal Sierpinski Gasket Fractal Patch Antenna consists of Antenna *A* and Antenna *B*, **a** Simulation in CST, **b** Fabricated version

5.72 GHz shows the second part. The return loss for both parts is shown as − 51.48 dB and − 37.94 dB. The targeted frequencies of the antenna at the 3.5 GHz and 5.8 GHz are shown as − 31.70 dB and − 22.00 dB, and this value is acceptable for the operate application of WiMAX and WLAN.

Table 2 shows the radiation pattern for Antenna *B*. The radiation pattern for Antenna *B* is shown in Table 2 at phi $= 0^0$ and phi $= 90^0$. At 3.5 GHz resonance frequency, it shows the same size (balanced shaped) for the main and also the back lobe of phi $= 0$. At phi $= 90$, it shows the same shaped with the different path of right and left.

For 5.8 GHz it has the same condition for phi $= 0$ with the equal same size of the forward and back lobes. For the phi $= 90$, the condition is different with the lobes which is focused on the right side (Fig. 4).

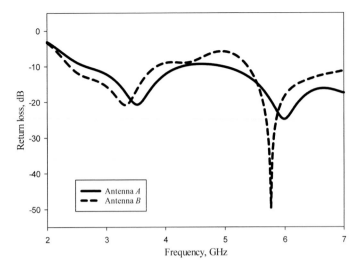

Fig. 3 Return loss of Antenna *A* and Antenna *B*

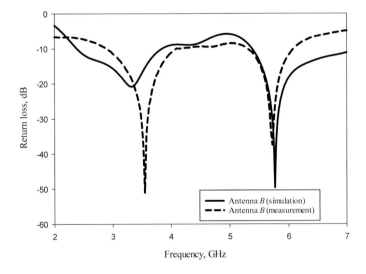

Fig. 4 Return loss of Antenna B (simulation and measurement)

Table 2 Radiation pattern for Antenna *B*

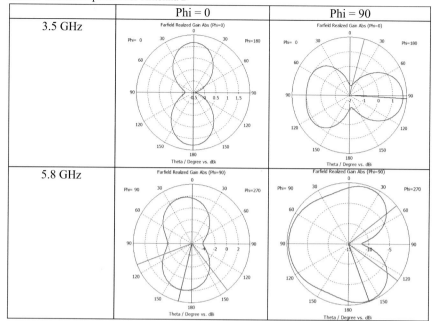

4 Conclusion

In this work, the initiator of Antenna *A* and the second iteration of Antenna *B* is done. From the investigations, it is observed that the change of the initiator shaped to the first iteration step of the pentagonal fractal had been effect to the location of the resonant frequency and also effect to improve the performance of the return loss of proposed antenna.

References

1. G.A. Deschamps, *Microstrip Microwave Antennas.* 3rd USAF Symposium on Antennas (1953)
2. T. Mondal, S. Suman, S. Singh, Novel design of fern fractal based triangular patch antenna. In: National conference on emerging trends on sustainable technology and engineering applications (NCETSTEA), pp. 1–3 (2020)
3. H. Yang, W. Yang, An ultra-wideband microstrip antenna based on Koch fractal resonance unit and CSRRs defective ground unit, In *9th Asia-Pacific Conference on Antennas and Propagation (APCAP)*, p. 1–2 ((2020))
4. B.B. Mandelbrot, *Fractals: Form, Chance and Dimension* (Freeman & Company, W. H. 1977)
5. B.B. Mandelbrot, *The Many Faces of Scaling: Fractals, Geometry of Nature, and Economics.* Self-Organization and Dissipative, pp. 91–109 (1982)
6. R. Goonatilake, R.A. Casas, Fractals, self-similarity, and beyond. Math. Teach. Res. J. **12**(1), 17–44 (2020)

7. R. Cicchetti, E. Miozzi, O. Testa, Wideband and UWB antennas for wireless applications: a comprehensive review. Int. J. Ant. Propagat. 1–45 (2017)
8. Y.B. Chaouche, M. Nedil, B. Hammache, M. Belazzoug, Design of modified sierpinski gasket fractal antenna for tri-band applications, in *IEEE International Symposium on Antennas and Propagation and USNC-URSI Radio Science Meeting*, pp. 889–890 (2019)
9. C.B. Nsir, J.-M. Ribero, C. Boussetta, A. Gharsallah, A wide band transparent Koch snowflake fractal antenna design for telecommunication applications, in *IEEE 19th Mediterranean Microwave Symposium (MMS)*, p. 1–3 (2019)
10. J. Raigoza, Differential private-Hilbert: data publication using Hilbert curve spatial mapping, in *International Conference on Computational Science and Computational Intelligence (CSCI)*, pp. 1493–1496 (2017)
11. B. Biswas, A. Karmakar, Design of ultra wide band monopole antenna with band rejection capability using minkowski fractal curve, in *IEEE Electrical Design of Advanced Packaging and Systems Symposium (EDAPS)*, pp. 1–3 (2018)
12. B. Benavides, R.A. Lituma, P.A. Chasi, L.F. Guerrero, A novel modified hexagonal shaped fractal antenna with multi band notch characteristics for UWB applications, in *IEEE-APS Topical Conference on Antennas and Propagation in Wireless Communications (APWC)*, pp. 830–833 (2018)
13. A. Gorai, M. Pal, R. Ghatak, A compact fractal-shaped antenna for ultrawideband and bluetooth wireless systems With WLAN rejection functionality. IEEE Antennas Wirel. Propag. Lett. **16**, 2163–2166 (2017)
14. V. Radonić, G. Misković, V. Cmojević-Bengin, *Fractal dipole antennas based on Hilbert curves with different line-to-spacing ratio*, in 13th International Conference on Advanced Technologies, Systems and Services in Telecommunications (TELSIKS), pp. 95–98 (2017)
15. W. Sierpiński, Sur une courbe cantorienne qui contient une image biunivoque et continue de toute courbe donnée. C. R. Acad. Sci. Paris. **162**, 629–632 (1916)
16. C. Puente, J. Romeu, R. Pous, X. Garcia, F. Benitez, Fractal multiband antenna based on the Sierpinski gasket. Electron. Lett. **32**(1), 1–2 (1996)

In-Pipe Navigation Robot for Nonuniform Pipe Diameter

Thivagar Kesavan, Mohamad Tarmizi Abu Seman, and Sattar Din

Abstract Pipeline inspection is a part of the pipeline integrity management for keeping the pipeline in good condition. The pipeline internal inspection is normally carried out through nondestructive testing techniques and technologies such as magnetic-flux leakage technology inn axial and circumferential, ultrasound technologies, eddy-current technologies and in-pipe inspection robot. Traditional in-pipe navigation robots are of rigid type with some limitations in terms of adaptability to different pipe diameters. This paper describes the development of a soft in-pipe navigation robot which can hop and crawl for maneuverability in horizontal and vertical pipes for pipe inspection. The robot consists of four motors, cables, elastic ribbons and cable-driven soft artificial muscles. The elastic ribbons control the contraction and elongation of the robot for propulsion, while the artificial muscles anchor the robot to provide a strong grip in a pipeline. The buckling and releasing characteristics of the artificial muscle and elastic ribbons are studied based on the performance of the robot. These findings can aid in navigations and inspection inside pipelines with corners and different pipe diameters.

Keywords Soft robot · In-pipe navigation robot · Artificial muscle

1 Introduction

Pipeline inspection is a part of the pipeline integrity management for keeping the pipeline in good condition Unfortunately, the cost of manual pipe inspection by human can be extremely high, especially with increasing complexity of pipeline nowadays which would require more manpower, and it is time consuming [1]. Thus,

T. Kesavan · M. T. A. Seman · S. Din (✉)
School of Electrical and Electronic Engineering, Universiti Sains Malaysia, 14300 Nibong Tebal, Penang, Malaysia
e-mail: sattar@usm.my

M. T. A. Seman
e-mail: mohdtarmizi@usm.my

pipe inspection robot is of great importance from both safety and cost perspectives. Therefore, various types of in-pipe robots have been developed for this purpose. The most well-known one is the pipeline inspection gauge, or pig, a passive data-collection device propelled by additionally applied high hydraulic pressure [2]. Traditional pipe inspection robots are of a rigid type, which have limitations in terms of adaptability to different pipe structures and dimensions. It has been reported that soft robotics has great potential to overcome the aforementioned limitations in in-pipe inspection [3].

A soft robotics is mainly constructed using stretchable, flexible materials such as silicon rubber [4] due to their low weight, high flexibility and large deformation [5]. In addition, the soft materials, which are the main component in artificial muscles of soft robots, enable the robot to perform various complex maneuvers such as contraction [6, 7], bending [6–9], twisting [7, 10, 11], helical notions [12, 13] while maintaining a good structural integrity. These softer robots are safer to use in hazardous workspace where human incapable to work such as in-pipe inspection where the space inside a pipe is small for human to reach [14, 15].

2 Methodology

2.1 Design

The soft in-pipe robot is based on a working principle of the earthworm-like locomotion for every forward and backward movements. In this work, the soft in-pipe robot has three subunits, i.e., the rear actuator, the middle actuators and the head actuator. The middle part consists of six elastic ribbons and four cables that connect to the two servomotors in head and rear actuators. The head and rear actuators contain four servomotors. Two servomotors control the buckling of the six elastic ribbons, while the other two servomotors control the contraction of six twisting-contraction artificial muscles (TCAM) [16]. Figure 1a, b show the cross-sectional view and 3D view of the CAD model of the robot.

The middle segment: To actuate the six elastic ribbons in the middle part, a pulley system was used to convert servomotor rotation in x-axis to z-axis motion. By using perimeter of circle Eq. 1 and torque of wheel Eq. 2, we can determine the radius of the servomotor horn which is 16.0 mm and the force produced by the horn which is 9375 g. Total rotation of servomotor per cycle = $(180°)/(360°) = 1/2$.

$$L = 2\pi r/2, \tag{1}$$

$$T = Fr, \tag{2}$$

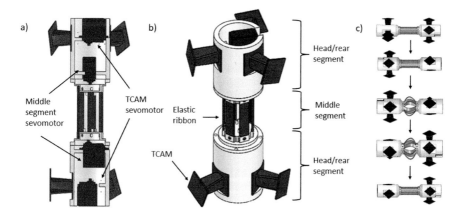

Fig. 1 **a** Cross-sectional view of the robot design; **b** 3D robot design; **c** sequence of motion to the left

where L is the length of nylon string to be displaced ($L= 50$ mm), r is the radius of the servomotor horn, T is the rated torque of the servo motor and F is the force produced by the servomotor.

The head and rear segment: For the head actuators and rear actuator, all the components are the same. Each segment contains two servomotors and three twisting-contraction artificial muscles (TCAMs). The TCAMs were first introduced in [16] as a form of a multipurpose artificial muscle. In their work, however, the TCAM was actuated using pneumatic system. In this work, on the other hand, the TCAMs are driven by cables and servomotors. Therefore, a series of fine tuning to the design is performed in this work. For this robot, the TCAMs are actuated using PE-braided nylon string, which is controlled by servomotors. Four nylon strings were used to tie in spiral connection with the TCAM to the four-side servomotor horn. The height of the TCAM was fixed at 40.0 mm, while the turning angle, wall thickness and edge thickness will be determined by conducting several experiments on the TCAM. Figure 2c shows the sequence of contraction and relaxation of TCAM and bending of elastic ribbon of the robot during the motion.

2.2 Fabrication

The fabrication of the in-pipe soft robot involves 3D printing and assembling. There are two different filaments which were used to fabricate the in-pipe soft robot. First, the polylactic acid (PLA) filament was used to print the case for the head and rear segments. Second, the thermoplastic polyurethane (TPU) filament was used to print TCAMs and elastic ribbons. Creality Ender-3 plus and Ender-5 plus were used 3D print these parts. Figure 2 shows the complete assembly of the fabricated robot.

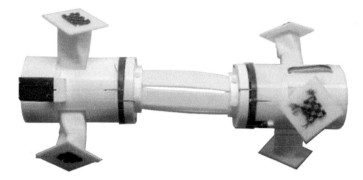

Fig. 2 Fabricated robot assembly

3 Results and Discussion

3.1 Validation of TCAM Model

A series of experiment was carried out to determine the optimum wall thickness, turning angle and edge thickness of the TCAM and the results are shown in Fig. 3a. Relaxation force at different contraction lengths will determine the force that the robot exerts on the inner wall of a pipe of different diameters. Based on the result, TCAM with wall thickness (T) of 1.2 mm, edge thickness (E) of 3.0 mm and tuning angle (A) of 45° produces highest relaxation force at different contraction lengths and therefore is chosen for the robot. From the result, it can be found that the more the TCAM contracts, the higher the force it can exert on the pipe wall.

Therefore, the in-pipe robot can generate more pushing force in small diameter pipeline. The TCAM can lift up to 4500 g of workload during vertical motion in a pipeline. From Fig. 3, we can see that TCAM with wall thickness of 1.2 mm, edge

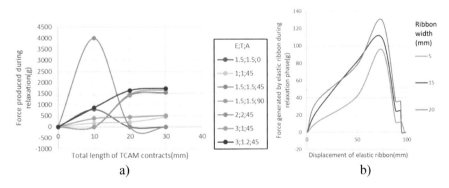

Fig. 3 **a** Force produced during relaxation versus contraction distance of TCAM; **b** force produced during relaxation versus bending distance of elastic ribbon

thickness of 3.0 mm and tuning angle of 45° can contract more than its predecessors and the force produced during relaxation is almost similar to TCAM with wall thickness of 1.5 mm which makes it the most suitable TCAM to navigate a soft in-pipe robot in nonuniform pipe diameter.

3.2   Validation of Elastic Ribbon

Next, a series of experiment was conducted to find the best parameters for the elastic ribbon and the results are shown in Fig. 3b. Based on the result, we understand that the more the elastic ribbon contracts, the higher the force it can generates. Therefore, in-pipe robot can generate more pushing force to move forward and backward. Hooke's law states that spring constant is directly proportional to cross-sectional area ($k \propto A$). So, by increasing the thickness of the elastic ribbon, the cross-sectional area of the elastic ribbon increases; thus, the spring constant of the elastic ribbon also increases. Therefore, the restoring force produced by the elastic ribbon increases. So, in order to produce a larger pushing force, the thickness of elastic ribbon must be higher.

Based on the experimental data, the final in-pipe soft robot was fabricated, which consists of six TCAMs with wall thickness of 1.2 mm, edge thickness of 3.0 mm and turning angle of 45°. The middle segment consists of six elastic ribbons with thickness of 3.0 mm, width of 15.0 mm and length of 80.0 mm. The final prototype is shown in Fig. 2.

3.3   Robot Navigation and Load-Carrying Capacity

To validate the performance of the final prototype of the robot, a series of experiment was conducted in which the robot needs to navigate in a straight pipe of different diameters and through a T-junction. The setup is shown in Fig. 4a, b. The results are summarized in Table 1. During moving in horizontal pipe, the speed of the in-pipe soft robot in the pipe with 100.0 mm diameter is higher than the speed of the in-pipe soft robot in the pipe with 84.0 mm because the elastic ribbons in middle segment contract more in the pipe with 100.0 mm diameter compared to pipe with 84.0 mm diameter. Thus, the in-pipe soft robot in 100.0 mm pipe diameter can cover more ground for every step compared to in-pipe soft robot in 84.0 mm pipe diameter so it moves faster in 100.0 mm pipe diameter. During moving in vertical pipe, the speed of the in-pipe soft robot in pipe with 100.0 mm diameter is lower than the speed of the in-pipe soft robot in pipe with 84.0 mm because the TCAMs skit more in pipe with diameter of 100.0 mm compared with pipe with diameter of 84.0 mm. This has validated the experimental results of TCAM which states that the more the TCAM contracts, the more the force it generates during deformation; therefore, coefficient of friction (COF) between the TCAM and acrylic pipe wall was higher which decreases the tendency of the in-pipe robot to skit.

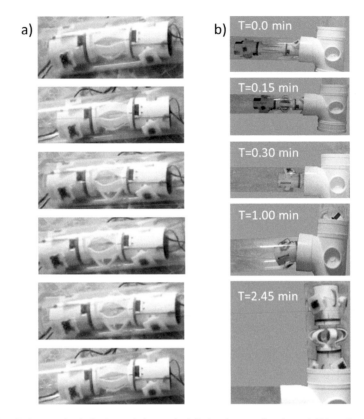

Fig. 4 **a** Robot moving in horizontal pipe to the left; **b** robot moving through T-junction

Table 1 Speed of the robot in horizontal and vertical pipes

Pipe diameter (mm)	Speed in horizontal pipe (mm/s)	Speed in vertical pipe (mm/s)
84	3.68	1.41
100	4.68	0.76

Another experiment was conducted to determine the maximum load-carrying capacity of the robot, and the results are summarized in Table 2. The maximum load the in-pipe soft robot can pull is similar for different pipe diameters which are 5000 g for horizontal pipe and 1373 g for vertical pipe because the pulling capacity depends on the torque of the servomotor. Therefore, the ability of the robot to pull the load can be improved by either reducing the robot weight or increasing the servomotor's torque. This shows that the in-pipe soft robot can pull more workload compared to the soft hopping and crawling robot [17]. Moreover, the maximum load the in-pipe soft robot can push in a 100.0 mm pipe diameter is higher than the maximum load the in-pipe soft robot can push in an 84.0 mm pipe diameter because the elastic ribbons

Table 2 Maximum weight that can be pulled and pushed by the robot in horizontal and vertical pipes

Pipe diameter (mm)	Horizontal pipe		Vertical pipe	
	Max. weight pulled (g)	Max. weight pushed (g)	Max. weight pulled (g)	Max. weight pushed (g)
84	5000	1338	1373	580
100	5000	2711	1373	728

in middle segment contract more in the pipe with 100.0 mm diameter compared to pipe with 84.0 mm diameter. This has validated the experimental results of the elastic ribbons which states that the more the elastic ribbon contracts, the more the force it generates during deformation; therefore, it can push more workload. Figure 4b shows the in-pipe soft robot crawling in the T-shaped pipe connector with inner diameter of 100.0 mm.

The in-pipe soft robot took around three minutes to navigate through the T-shaped PVC pipe connector. This is because the hard PLA parts of the robot cannot bend; therefore, the robot needs to turn right by 10° for every cycle of forward movement until it reaches 90° angle of turning. Thus, the time taken for in-pipe soft robot to navigate in a T-shaped pipe connector increases.

4 Conclusion

In this project, we presented an in-pipe soft robot to navigate in a nonuniform pipe diameter. Our robot has excellent performances: During horizontal crawling, it reaches a maximum speed 4.68 mm/s when crawling a horizontal pipe and 1.41 mm/s when crawling vertical pipe. It also has a good load-carrying capability. In a horizontal pipe, it is able to push 4.23-fold and pull 7.80-fold of its weight, while push 1.33-fold and pull 2.14-fold of its weight in a vertical pipe. Even though it has hard PLA parts, the in-pipe soft robot was able to bend and crawl through pipe connector with 90° turns. In crawling test, the in-pipe soft robot was also crawled through pipes with diameter changing from 84.0 to 100.0 mm. In the future, Central Pattern Generators (CPGs) will be implemented into the in-pipe soft robot locomotion.

Acknowledgements This work was funded by the School of Electrical and Electronics Engineering, Universiti Sains Malaysia via the short-term grant 304/PELECT/6315477.

References

1. Y. Bai, Q. Bai, *Subsea Pipeline Integrity and Risk Management* (Gulf Professional Publishing, 2014)
2. M. Ho, S. El-Borgi, D. Patil, G. Song, Inspection and monitoring systems subsea pipelines: a review paper. Struc Health Monit **19**(2), 606–645 (2020)
3. A. Shiva, A. Stilli, Y. Noh, A. Faragasso, I. De Falco, G. Gerboni, H.A. Wurdemann, Tendon-based stiffening for a pneumatically actuated soft manipulator. IEEE Robot Autom. Lett. **1**(2), 632–637 (2016)
4. S.C. Shit, P. Shah, A review on silicone rubber. Natl. Acad. Sci. Lett. **36**(4), 355–365 (2013)
5. T. Aziz, M. Waters, R. Jagger, Analysis of the properties of silicone rubber maxillofacial prosthetic materials. J. Dent. **31**(1), 67–74 (2003)
6. J. Li, T. Biel, P. Lomada, Q. Yu, T. Kim, Buckling-induced F-actin fragmentation modulates the contraction of active cytoskeletal networks. Soft Matter **13**(17), 3213–3220 (2017)
7. M. Schaffner, J.A. Faber, L. Pianegonda, P.A. Rühs, F. Coulter, A.R. Studart, 3D printing of robotic soft actuators with programmable bioinspired architectures. Nat. Commun. **9**(1), 1–9 (2018)
8. Q. Zhang, W. Wang, J. Zhang, X. Zhu, L. Fu, Thermally induced bending of ReS2 nanowalls. Adv. Mater. **30**(3), 1704585 (2018)
9. P. Xiao, N. Yi, T. Zhang, Y. Huang, H. Chang, Y. Yang, Y. Chen, Construction of a fish-like robot based on high performance graphene/PVDF bimorph actuation materials. Adv. Sci. **3**(6), 1500438 (2016)
10. D. Yang, B. Mosadegh, A. Ainla, B. Lee, F. Khashai, Z. Suo, G.M. Whitesides, Buckling of elastomeric beams enables actuation of soft machines. Adv. Mater. **27**(41), 6323–6327 (2015)
11. J. Foroughi, G.M. Spinks, G.G. Wallace, J. Oh, M.E. Kozlov, S. Fang, R.H. Baughman, Torsional carbon nanotube artificial muscles. Science **334**(6055), 494–497 (2011)
12. F. Zhang, L. Xiong, Y. Ai, Z. Liang, Q. Liang, Stretchable multiresponsive hydrogel with actuatable, shape memory, and self-healing properties. Adv. Sci. **5**(8), 1800450 (2018)
13. R.V. Martinez, J.L. Branch, C.R. Fish, L. Jin, R.F. Shepherd, R.M. Nunes, G.M. Whitesides, Robotic tentacles with three-dimensional mobility based on flexible elastomers. Adv. Mater. **25**(2), 205–212 (2013)
14. X. Zhang, T. Pan, H.L. Heung, P.W.Y. Chiu, Z. Li, A biomimetic soft robot for inspecting pipeline with significant diameter variation, in *2018 IEEE/RSJ International Conference on Intelligent Robots and Systems* (2018), pp. 7486–749
15. M.S. Verma, A. Ainla, D. Yang, D. Harburg, G.M. Whitesides, A soft tube-climbing robot. Soft Robot. **5**(2), 133–137 (2018)
16. Z. Jiao, C. Zhang, W. Wang, M. Pan, H. Yang, J. Zou, Advanced artificial muscle for flexible material-based reconfigurable soft robots. Adv. Sci. **6**(21), 1901371 (2019)
17. C.Y. Yeh, C.Y. Chen, J.Y. Juang, Soft hopping and crawling robot for in-pipe traveling. Extreme Mech. Lett. **39**, 100854 (2020)

Development of a Health Indication System via Vibration for Pre-diagnosis

Mohd Fauzi Abu Hassan, Muhammad Nazrul Zharif Mohd Zaini, Muhammad Khusairi Osman, Zakiah Ahmad, and Zainal Nazri Mohd Yusuf

Abstract A health indication system device for machines is rarely produced as the condition of the machine usually checked by the machinist via the usual method and it is manually conducted. The usual method of determining the condition of the machine through vibration is quite complicated as it involves formulas and calculations. The issue of machine health is very paramount to any company that is involved in production works as the maintenance of the machines plays an important part to sustain the production quantity and quality. Therefore, a machine health indication system has been developed to indicate the condition of the machine through vibration checking confirming the actual condition or any changing machine's condition from time to time via a device. The machine health check is importantly required to avoid the development of minor problems into major problems which can lead into cost issues. In addition, the benchmark of this project is based on the process of fault severity determination on checking and monitoring the machine condition through

M. F. A. Hassan (✉) · M. N. Z. M. Zaini
Intelligent Automotive Systems Research Cluster, Electrical, Electronic and Automation Section, Malaysian Spanish Institute, Universiti Kuala Lumpur, Kulim Hi-Tech Park, 09000 Kulim, Kedah, Malaysia
e-mail: mohdfauzi@unikl.edu.my

M. N. Z. M. Zaini
e-mail: nazrul@s.unikl.edu.my

M. K. Osman
Faculty of Electrical Engineering, Universiti Teknologi MARA (UiTM), Cawangan Pulau Pinang, Kepala Batas, Malaysia
e-mail: khusairi@uitm.edu.my

Z. Ahmad
Engineering Section, Malaysian Spanish Institute, Universiti Kuala Lumpur, Kulim Hi-Tech Park, 09000 Kulim, Kedah, Malaysia
e-mail: zakiah@unikl.edu.my

Z. N. M. Yusuf
Mechanical Section, Malaysian Spanish Institute, Universiti Kuala Lumpur, Kulim Hi-Tech Park, 09000 Kulim, Kedah, Malaysia
e-mail: zainalnazri@unikl.edu.my

M. H. Abu Bakar et al. (eds.), *IT Solutions for Sustainable Living*,
SpringerBriefs in Applied Sciences and Technology,
https://doi.org/10.1007/978-3-031-51859-1_12

vibration as in the subject of vibration and analysis. The purpose of this project is to determine the condition of the machine through vibration in the fastest and simplest way possible. The use of a microcontroller, liquid crystal display, accelerometer, and some other electronic equipment assembled together help in the accomplishment of that. A mechanism or device based on the development of a health indication system for machines will be attached to the main motor of the machine near the motor bearing in order to attain the exact vibration value in determining the condition of the machine.

Keywords Health indication system · Condition monitoring · Vibration analysis · Fault severity determination · Machine condition

1 Introduction

A health indication system is a system used to determine or indicate the condition of health. This project is focusing on indicating the health condition of machines through the main part which is the motor. The condition needs to be determined as a method of pre-diagnosis in order to confirm the actual condition of the machine before it undergoes the maintenance process. In the topic of fault severity determination by the course manual of vibration analyst certification provided by the Institute of Noise and Vibrations, Universiti Teknologi Malaysia, stated that vibrations in practice are assessed for potential structure damage, machinery acceptance, the sensitivity of equipment and process to excessive vibration. Thus, this project will take the opportunity to be developed in order to indicate the condition of the machine's motor based on the vibration standard.

1.1 Problem Statement

The rotary machines in the industry are already under proper observation and care of the specialist. Common people with no knowledge in engineering find it hard to take care of their own machines as the problem can only be determined with high and proper engineering skills. They usually appoint the problem of their machines to the specialist once the machine is already in the worst condition or cannot be run anymore. Thus, the maintenance process is involved with cost issues. Things might be great for the common user to take care of their machines under their observation with their own maintenance check. This will help the machines to smoothly run for a long time as the minor problem can be detected and action can immediately be taken before it turns into a major problem that can bring the machine into the worst condition.

Besides that, fault severity determination is the current solution used by the experts to determine the condition of machines via vibration. The current solution for this problem is quite complicated for common users. A vibration meter can be used to define the condition of the motor, but the user needs to refer the values gained from the vibration meter through vibration analysis via vibration severity chart in order to obtain the actual condition of the motor. Moreover, the current solution is also involved with calculation and values conversion, so it will be very complicated and hard for common users to determine the motor condition. Thus, this project will be developed based on the current solution. By comparing to the current solution, the device output will directly show the results and condition of the motor through a display without any difficulty.

1.2 Objective

Throughout this project, the following objectives will be achieved:

1. To develop a health indication system that can indicate the condition of a machine.
2. To identify any significant changes which are indicative for the generate of any faults in the machine.
3. To provide a low-cost product that can be effective in determining the condition of the machine through vibration analysis.

2 Literature

The development of this project is related to the idea, sources, and articles available on online platforms. There was a lot of research conducted that helps to provide an improvement for the project to be more effective and convenient. Furthermore, there are a lot of sources about technology that can be used to enhance the ability of the project. To achieve the objectives, the literature review is very important. The review of other projects is worth on purpose of gaining knowledge and skills needed while completing this project. Sources from other theses which proved to be a useful are taken in identifying problems and give improvements for the analysis and decision-making in this project.

2.1 Overview of Condition Monitoring Technology

Condition monitoring is the method of monitoring a condition parameter in a vibration machinery [1–8] to assess a significant change that is indicative of a failure to grow. It is a major component of predictive maintenance. Using condition monitoring allows for the scheduling of maintenance or other actions to be taken to prevent consequential damage and to avoid its consequences. Condition monitoring has a significant benefit since symptoms can be resolved before they grow into a major failure which would shorten the normal lifespan. In addition, condition monitoring methods are usually used on rotating equipment, auxiliary systems, and other types of machinery [7, 8] such as compressors, generators, electric motors, internal combustion engines, and presses, while on static plant equipment such as steam boilers, piping, and heat exchangers, regular inspection using non-destructive testing (NDT) methods and fit for service (FFS) evaluation is used.

2.2 *Overview of Vibration Analysis*

Vibration is the movement or mechanical oscillation surrounding a machine or component's equilibrium position [9–11]. This can be repetitive, like a pendulum motion, or random, like a tire rolling on a gravel track. Vibration can be expressed in metric units (m/s^2) or gravitational constant units "g", where 1 g = 9.81 m/s^2 is used. An object can vibrate in two ways which are free vibration and forced vibration. Free vibration happens when an object or structure is displaced or impacted and then allows for natural oscillation. Based on the journal paper "Measuring mechanical vibrations using an Arduino as a slave I/O to an EPICS control system", it is stated that the hardware and software were designed to test mechanical vibrations in the FREIA laboratory at Uppsala University [6]. They used an Arduino microcontroller as a slave I/O and fitted it with dual accelerometers to be used for measuring vibrations and a serial adapter to link the hardware to an EPICS IOC for analysis. In order to find a transfer function, data from the two accelerometers was then cross-correlated. The results were in good theoretical agreement. When designing physical experiments, it is of utmost importance that one takes into account the mechanical movements that occur and influence the results. There are many ways to quantify mechanical vibrations, and in this research, they are looking more closely at how to quantify them using an accelerometer based on MEMS, ADXL 335. Vibration transfers from one point to another can be seen by using two accelerometers and thereby gain some information about the characteristics of the medium through which the vibrations propagated. An Arduino microcontroller is used to provide power to the accelerometers and to capture the waveforms.

3   Methodology

The development of this final year project was conducted in the Vibration Laboratory, Fluid Mechanics Laboratory, and the chiller motor storage room in the Universiti Kuala Lumpur Malaysian Spanish Institute, Hi-Tech Park, Kulim, Kedah. The study on vibration analysis through fault severity determination was conducted based on the lab task of condition monitoring by the subject of vibration and noise, SCB35403. The purpose of the experiment is to study the effect of machine condition through vibration analysis. Furthermore, the attempt to collect data and testing the project has

Fig. 1 Process flowchart

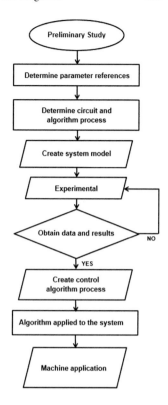

been done on a machine in the Fluid Mechanics Laboratory. A vibration meter with an accelerometer sensor has been used as the benchmark to be compared with the developed project. On top of that, the whole process of collecting the actual data on vibration analysis was conducted in the motor storage room for chiller system based on two AC motors with different conditions. The development of this project includes various phases such as preliminary study, experimental investigations, determination of parameter references, determination of circuit and algorithm process, creation of control algorithm and system model, then obtaining data and results, and finally applies the complete system to a machine. There are several information and research needed to be linked with this project to meet the requirements of this project as stated in Fig. 1. All information regarding project operations, circuits, and applications have been studied in this project.

3.1 Study on Condition Monitoring

Condition monitoring is a method used to monitor the condition of machine in vibration analysis. The study on this topic was conducted in the Vibration Laboratory and

Fluid Mechanics Laboratory. In addition, the study has been done by the guidance of co-supervisor and technician in charge. The application of this study is necessary for the development of the health indication system project as it is related to the whole process of the system requirements. The study starts by referring to the course manual of vibration analyst certification by the Institute of Noise and Vibrations, Universiti Teknikal Malaysia, on the topic fault severity determination. A practical task was conducted in order to understand the concept of condition monitoring and fault severity determination. The title for the practical task is condition monitoring, conducted by the subject vibration and noise, SCB35403. The first objective of the task is to measure the vibration amplitude in displacement, velocity, and acceleration by using a vibration meter. The second objective is to determine the condition of the vibrating system using a vibration severity chart. A vibration meter with accelerometer sensor is used to analyze the vibration occurring at the bearings of the motor. Based on the condition monitoring study, the condition of the machine can be determined through a severity chart. As for this practical task, the focus is on the value of vibration in velocity unit and the rotating speed of the testing motor. The condition of the machine can exactly be determined by plotting the velocity value against the rotating speed of the motor.

3.2   Application of Alternating Current (AC) Motor

The application of an alternating current (AC) motor has been appointed through this project development. Two AC motors with the same specification have been applied for the development of this project in order to achieve the objectives of this final year project. The experiment was conducted in the storage room of motors for chiller systems in the Universiti Kuala Lumpur Malaysian Spanish Institute. There are four AC motors in the room, and only two of them have been applied for this project as the two motors run with different conditions. Moreover, a vibration meter with accelerometer sensor has been used to determine the condition of the motor as reference solution before the current project has tested the motors.

In this experiment, the data is being collected based on two AC motors as they are determined to be run with different conditions which are good and bad. This project is being developed with an Arduino microcontroller and the data collected through the system is determined by an accelerometer sensor applied to the casing of the motor near to the main bearings of the two motors. Three experiments have been conducted through this study in order to confirm the exact conditions produced by both motors. The data taken from all three experiments of the study is recorded for further development of the system model. Figure 2 shows the reading on vibration analysis of the motor which has been monitored through the software. The data from the monitor is then transferred into Microsoft Excel. From the data required, the vibration and noise between the two motors can be confirmed.

a  b

Fig. 2 **a** AC motors used for vibration analysis. **b** Plotter monitor on Arduino IDE

3.3 Hardware Development

In order to achieve the objectives of this project, there are a few hardware pieces that have been involved. The hardware development for this project consists of the microcontroller, liquid crystal display, AC motors, and other few materials. Thus, for the application of hardware development through this project, every single hardware and component features applied to this project are described as follows:

1. Castle GA2002 vibration meter and sensor.
2. Arduino UNO Rev3 microcontroller.
3. ADXL 345 MEMS accelerometer.
4. I2C 1602 character LCD display module.
5. Breadboard.
6. Magnetic plate.
7. Jumper wire.
8. Piezo buzzer.
9. Light-emitting diode (LED).

3.4 Software Development

The software development has the main role to ensure that all the operations of the system are working since the system requires the use of the Arduino microcontroller. Moreover, software is the term used for the instruction to inform the Arduino what to do with all the received data in a certain task. In terms of hardware and on uploading the code to the main component is needed for compiling. The program is written and assembled to the Arduino microcontroller by using the C language. The C languages are one of the high-level languages for programming computer, microprocessor, and microcontroller. In addition, the Tera Term software and Microsoft Excel are being used to collect the data received from the microcontroller. The software used for the development of this project is as follows.

1. Arduino IDE.
2. Tera Term.
3. Microsoft Excel.

4   Result and Discussion

The condition monitoring study by doing the practical task, referring to the subject of vibration and noise, provided raw data that can be taken as one of the methods to completely make this project become successful. Table 1 shows the exact data that can be achieved during the practical task in order to understand the process of fault severity determination. In this study, the best value taken to determine the condition of motor is the vibration velocity. Based on Table 1, data has been taken in determining the exact condition of the motor on the front and rear positions of the motor bearings at different motor speeds. A vibration meter has been used in conducting this process of collecting the data based on the lab task, condition monitoring. Table 1 shows three (3) kind of parameters were taken from the experiment. As for this project, the method is focusing on the velocity as it is representing the vibration value from the motor bearings. In addition, during this experiment, every single position with different speeds shows that the motor is in good condition. The data taken has been referred to the severity chart before confirming the exact motor condition. Thus, the overall process of this experiment and the condition monitoring method has shown that the motor used in this study is in good and smooth condition.

Condition monitoring is the most important method in the accomplishment of this project. The same method as used in the study of condition monitoring has been applied to two AC motors. The value gained by the vibration meter is set as the benchmark for this project to be compared. The data achieved from this experiment is set to be used as the algorithm for the system model. The process to analyze the data has been described as in the previous chapter. Thus, data shown in Table 1 represents the vibration measurement gained from the project system model. There are two AC motors involved that represent two different conditions which are smooth and rough. The condition of motors has been confirmed first by the vibration meter as the value is being referred to the vibration severity chart. Furthermore, three experiments have

Table 1 Condition monitoring data

Speed (rpm)	Position	Displacement (μm)	Velocity (m/s)	Acceleration (m/s^2)	Condition
700	Front	0.008	0.22	1.19	Good
700	Rear	0.008	0.26	2.32	Good
900	Front	0.003	0.18	1.46	Good
900	Rear	0.004	0.39	3.34	Good
1100	Front	0.005	0.23	1.75	Good
1100	Rear	0.003	1.53	1.53	Good

been done as for confirming the exact condition of the motors before the data taken is developed into the system model of this project. As for this experiment, the condition monitoring method presents the data in the form of velocity against rotation speed of the motor in order to show the condition of the vibration produced by the two AC motors based on the severity chart.

All of the data taken from Experiment 1, Experiment 2, and Experiment 3 have been combined together in a graph to show the exact condition of the two AC motors. As seen in Fig. 3, AC motor A and AC motor B are presented with different conditions based on the severity chart and condition zones. Moreover, AC motor A has finally been verified to be in the state of smooth condition as the three plots for the data taken from the three experiments before show that the velocity of the vibration presented is in the state of smooth condition. Meanwhile, as for AC motor B, the three plots for the data taken from the experiments before show that the condition of the motor is in the state of rough. Thus, this data is successfully taken and fixed to be set as the reference for the project model. An analysis based on the data taken for the vibration analysis on the AC motors has been done via the system model for this project. The data taken has been set up as the benchmark for the model to determine the exact and same condition for the motors via the condition monitoring method. The health indication system receives the vibration signal in the form of amplitude against the sample time. As for this project, the system model presents the data in waveform of amplitude against time in order to show the pattern and signal of the vibration produced by the two AC motors. Thus, as the process is focusing on two different conditions, the project model received two different signals that represent the condition which are smooth condition and rough condition. The analysis of the vibration is focusing on two different motors that will affect the condition and health of the machines. Thus, the measurement of the vibration required from AC motor A has been confirmed to represent the smooth condition and AC motor B as rough condition. The signals received by the motor have been transformed into waveform of amplitude against the number of samples taken. Three hundred samples are taken for the measurement of the amplitude produced by the vibration from the motor.

4.1 Smooth Condition

The data required from the waveform of the amplitude against time as in Fig. 4 shows that the highest peak obtained from the smooth condition by the health indication system is 3.77 m/s^2, while the lowest is -2.2 m/s^2. As seen by the waveform, this data needs its standard deviation so that the actual range and average for this condition can be set by the algorithm and the normal distribution curve can be constructed. In order to define its standard deviation, the data formed by the health indication system needs to undergo the formula as below:

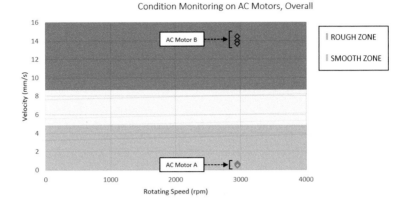

Fig. 3 Condition monitoring data on AC motor for all experiments

$$\sigma = \sqrt{\frac{\sum (x_i - \mu)^2}{N}}$$

By determined the average and variance, the standard deviation of the data formed by the smooth condition can be obtained. The standard deviation and other required data are stated in Table 2.

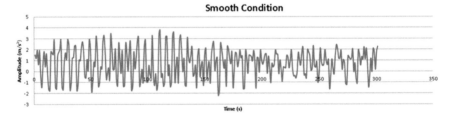

Fig. 4 Waveform of smooth condition

Table 2 Standard deviation on smooth data

Data	Value
Sum, Σ	217.07
Count, n	300
Mean, μ	0.723567
Variance, σ^2	1.843109
Standard deviation, σ	1.357611
$\mu - \sigma$	−0.63404
$\mu + \sigma$	2.081178

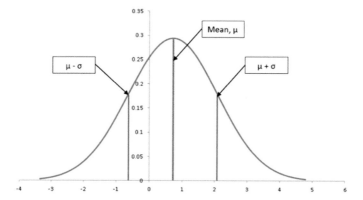

Fig. 5 Normal distribution curve of smooth condition

In Table 2, the sum of data formed by the health indication system is 217.03 m/s². From the sum of all the values by the data, the mean can be obtained by dividing the sum value with the value of sample which is the count. The standard deviation can finally be recorded over the variance. In addition, the mean value then will be subtracted from and added to the standard deviation to define the optional range for the smooth condition. From the data above, a normal distribution curve for this condition can be formed as shown in Fig. 5.

Figure 5 shows the normal distribution curve data is presented. The mean line of the data is shown in green color which means the average of all values obtained from the data analysis. As for the mean, the value can be retrieved as 0.723567. In other way, the two red lines on the normal distribution curve data represent the exact range of vibration frequency obtained from AC motor A. Furthermore, the range of the vibration frequency can be obtained by adding and subtracting the value of mean with the standard deviation. In this form, the range of the vibration frequency is in between −0.63404 and 2.081178. From the normal distribution curve achieved based on the standard deviation of the data acquired from the vibration analysis, the algorithm for this project can also be achieved.

4.2 Rough Condition

The data required from the waveform of the amplitude against time as in Fig. 6 shows that the highest peak obtained from the smooth condition by the health indication system is 3.77 m/s² while the lowest is −2.2 m/s². As seen by the waveform, this data needs its standard deviation so that the actual range and average for this condition can be set by the algorithm and the normal distribution curve can be constructed. In order to define its standard deviation, the data formed by the health indication system needs to undergo the same formula (1). By determined the average and variance, standard

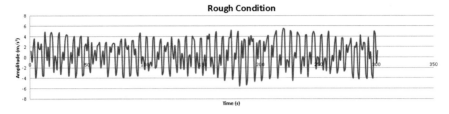

Fig. 6 Waveform of rough condition

Table 3 Standard deviation on rough data

Data	Value
Sum, Σ	185.17
Count, n	300
Mean, μ	0.617233
Variance, σ^2	8.156443
Standard deviation, σ	2.855949
$\mu - \sigma$	−2.23872
$\mu + \sigma$	3.473182

deviation of the data formed by the rough condition can be obtained. The standard deviation and other required data are stated in Table 3.

In Table 3, the sum of data formed by the health indication system is 185.17 m/s². From the sum of all values by the data, the mean can be obtained by dividing the sum value with the value of sample which is the count. The standard deviation can finally be recorded over the variance. In addition, the mean value then will be subtracted and added with standard deviation to define the optional range for smooth condition. From the data above, the normal distribution curve for this condition can be formed as shown in Fig. 7.

Based on Fig. 7, the normal distribution curve data is presented the same as in the smooth condition. The mean line of the data is also shown in green color which means the average of the all values obtained from the data analysis by AC motor B. As for the mean, the value that can be retrieved is 0.617233. In other way, the two red lines on the normal distribution curve data represent the exact range of vibration frequency obtained from AC motor B. In addition, the range of the vibration frequency can be obtained by adding and subtracting the value of mean with the standard deviation. In this form, the range of the vibration frequency for AC motor B is in between −2.23872 and 3.473182. From the normal distribution curve achieved based on the standard deviation of the data acquired from the vibration analysis, the algorithm for this project can also be achieved.

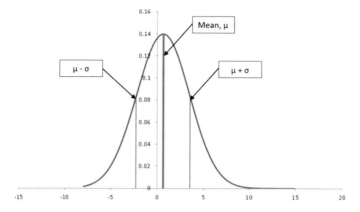

Fig. 7 Normal distribution curve of rough condition

4.3 Smooth and Rough Condition

The measurement of the vibration required from both AC motors has been compiled in one graph. The analysis of the vibration is focusing on two different motors which are the AC motor A and AC motor B that will affect the condition and health of the machines. This is the way of method in order to show the comparison of the normal distribution curve data between the smooth condition and rough condition. The two data of the condition are shown in the waveform of amplitude against the count of 300 samples taken in time. Figure 8 shows amplitude comparison between smooth and rough condition.

Based on Fig. 8, the waveforms acquired by both conditions retrieved the signal of the two conditions that show the amplitude for smooth condition which is smaller than the rough condition. Both of the data were taken in the count of 300 samples in time. The signal of smooth condition is in blue color, while the condition of rough is in red color. Moreover, the comparison between the two conditions of smooth and rough can be seen in this state. The size of the amplitude between the two is different as the rough condition shows bigger size of amplitude than the condition of smooth.

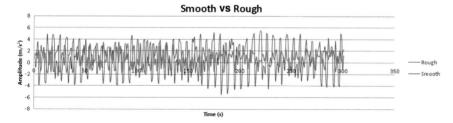

Fig. 8 Waveform of smooth and rough conditions

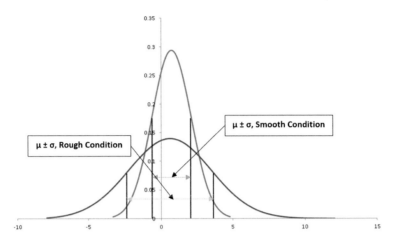

Fig. 9 Normal distribution curve of smooth and rough conditions

In addition, as the combination signal of the two conditions can be formed, the normal distribution curve for these two conditions can also be achieved (see Fig. 9).

Based on Fig. 9, the comparison of the normal distribution curve between the states of two conditions is formed as the blue line represents the smooth condition, while the red line represents the condition of rough. In this state, lines have been formed in the data in order to determine the exact details of the normal distribution curve of the two conditions. Thus, green lines formed in black color stated for the range of both conditions based on their mean and standard deviation values. In addition, as for the smooth condition, the range of vibration frequency obtained by calculation is in between −0.63404 and 2.081178, while for the condition of rough, the range obtained is in between −2.23872 and 3.473182. By achieving the normal distribution of both conditions, the algorithm from this analysis can also be achieved.

The algorithm retrieved based on the analysis between the two conditions is still not enough as the algorithm turned into a quite complicated state. Thus, another form of normal distribution curved in more details is present (see Fig. 10). The data achieved by the comparison of two conditions of smooth and rough in normal distribution curve state leads into the confirmation of the border between these two conditions. Besides that, the conditions of the motor need to be present in the exact state as in the fault severity chart. There must be a border that divide the vibration; smooth and rough condition. Thus, as in Fig. 10, the border that determines the zone of smooth condition and rough condition can be required based on the range values achieved for the normal distribution data of the conditions.

In this state of comparison between the two conditions, the border value that will be set as new range for the algorithm is in between −1.43638 and 2.77718. Thus, this means that the value that surpasses the border value will definitely be defined as rough condition, while the value that did not surpass will be defined as in the condition of smooth state. To be exact, the actual algorithm that can be used in the system model is as follows:

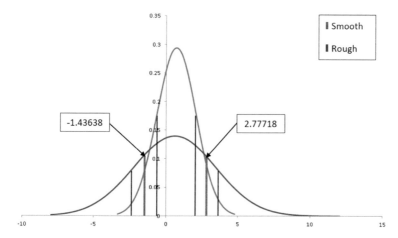

Fig. 10 Normal distribution curve of smooth and rough conditions in detail

IF vibration value is in between −1.43638 and 2.77718,
THEN the condition is SMOOTH,
ELSE the condition is ROUGH.

5 Conclusions

In the conclusion, the vibration analysis can be stated as one of the most important and basic methods for fault severity determination by machines in order to indicate the condition. The major purpose of fault severity determination is to detect possible upcoming damage to the machine and for the machine acceptance. The progress of developing this project has been conducted in order to prove that vibration analysis is the perfect method to determine the condition of the machine, so that immediate action can be taken to the machine if there is any condition change detected by the health indication system. The purpose is to avoid worst condition from occurring to the machine. In addition, there were limitations obtained on doing this project. The data provided by Table 1 is not perfectly equal to the benchmark's data as it is slightly different because of the quality of the sensor. But the data differences between the actual project and the benchmark are not that large. Thus, several tests between the two have been conducted in order to get the actual data that will give the same output at the end of the process. Luckily, at the end of the experiment, this project and the benchmark developed the same output through the condition of the machine's motor. Last but not least, based on the result it shows that the experiment has successfully increased. This means that the output of the final year project has been successfully achieved. Nevertheless, due to 2020 Movement Control Order (MCO) announced by the Government of Malaysia, the development plan on this project is becoming limited. Thus, a new development on this project with application of the Internet of

Things (IoT) will be an advanced system model, so the system works as a monitoring system as well. To summarize the whole things up, it is finally proven that vibration analysis is the best method in order to indicate the condition of the machine. Thus, with proper self-maintenance check on the machine, the lifetime of the equipment will greatly become extended for a long time as action can immediately be taken before the condition of the machine getting worst. The health indication system will help to sustain the quality and quantity of production works by the machine.

References

1. L. Cui, Z. Jin, J. Huang, H. Wang, Fault severity classification and size estimation for ball bearings based on vibration mechanism. Spec. Sect. Adv. Progn. Syst. Health Manag. **7**, 56107–56116 (2019)
2. N. Sawalhi, R.B. Randall, Vibration response of spalled rolling element bearings: observations, simulations and signal processing techniques to track the spall size. Mech. Syst. Sig. Proc. **25**(3), 846–870 (2011)
3. L. Song, P. Chen, H. Wang, Vibration-based intelligent fault diagnosis for roller bearings in low-speed rotating machinery. IEEE Trans. Instrum. Meas. **67**(8), 1887–1899 (2018)
4. M.K. Pradhan, P. Gupta, Fault detection using vibration signal analysis of rolling element bearing in time domain using an innovative time scalar indicator. Int. J. Manuf. Res. **12**(3), 305–317 (2017)
5. J. Mathew, R. Alfredson, The condition monitoring of rolling element bearings using vibration analysis. J. Vib. Accoust. **105**(3), 447–453 (1984)
6. Hjort A, Holmberg M (2015) Measuring mechanical vibrations using an arduino as a slave I/O to an epics control system. In: Department of Physics and Astronomy Uppsala University. Retrieved from http://uu.diva-portal.org
7. J. Yu, Y. Xu, G. Yu, L. Liu, Fault severity identification of roller bearings using flow graph and non-naive Bayesian inference. Proc. Inst. Mech. Eng. Part C **233**(14), 5161–5171 (2019)
8. Z. Zhang, M. Entezami, E. Stewart, C. Roberts, Enhanced fault diagnosis of roller bearing elements using a combination of empirical mode decomposition and minimum entropy deconvolution. Proc. Inst. Mech. Eng. Part C **231**(4), 655–671 (2017)
9. Chatterton S, Ricci R & Pennacchi P (2014) Signal processing diagnostic tool for rolling element bearings using EMD and MED. Advances in condition monitoring of machinery in non-stationary operations. Lect. Notes Mech. Eng. 379–388
10. Y. Ohue, A. Yoshida, M. Seki, Application of the wavelet transform to health monitoring and evaluation of dynamic characteristics in gear sets. Proc. Inst. Mech. Eng. Part. J. **218**(1), 1–11 (2004)
11. C.G. Harris, J.H. Williams, A. Davies, Condition monitoring of machine tools. Int. J. Prod. Res. **2**(9), 1445–1464 (1989)

Methods and Applications in Fluid Structure Interaction (FSI)

Khairul Shahril, Khairul Akmal, Shahril Nizam, Muhammad Najib, and Ishak Azid

Abstract The interaction of an internal or external fluid flow, such as aerodynamic or turbine flow, with any moveable or deformable structure is known as fluid structure interaction (FSI). Many industrial engineering challenges include fluid structure interaction analysis and dynamics analysis to increase functionality or operational efficiency. FSI is the framework behind the most elegant approaches developed in the FSI field by researchers. There are various methods by which the interaction analysis of the fluid structure can be performed. The current study's goal is to provide an overview of the importance of FSI in any context. Fluid structure interactions are a major concern in the design of many engineering systems, and failing to comprehend the consequences of oscillatory interactions, especially in material-based structures, can be fatal. The interaction of a fluid and a solid body happens all the time in nature, and it happens at all scales and disciplines. While one of the oldest and most classical problems in fluid mechanics is the mathematical theory of body motion in a liquid, mathematicians have only recently become interested in a systematic analysis of the fundamental problems related to fluid structure interaction from both theoretical and computational perspectives.

Keywords Fluid structure interaction · Turbine · Motion of bodies · Fluid mechanics · Fluid flow

K. Shahril (✉) · K. Akmal · S. Nizam · M. Najib · I. Azid
Universiti Kuala Lumpur, Malaysian Spanish Institute Kulim Hi-Tech Park, 09000 Kulim, Kedah, Malaysia
e-mail: khairuls@unikl.edu.my

K. Akmal
e-mail: khairulakmal@unikl.edu.my

S. Nizam
e-mail: shahrilnizam@unikl.edu.my

M. Najib
e-mail: mnajib@unikl.edu.my

I. Azid
e-mail: ishak.abdulazid@unikl.edu.my

© The Author(s), under exclusive license to Springer Nature Switzerland AG 2024
M. H. Abu Bakar et al. (eds.), *IT Solutions for Sustainable Living*,
SpringerBriefs in Applied Sciences and Technology,
https://doi.org/10.1007/978-3-031-51859-1_13

1 Introduction

Fluid structure interaction is an interdisciplinary subject of interest for many academics in the realm of fluid dynamics. The finite-element method has been at the forefront of research in this important subject. Fluid structure interaction occurs in both natural systems and man-made artefacts in many forms. The interaction of a tree and the wind, as well as groundwater contact with the soil, is typical instance of fluid structure interaction in nature. For engineered structures, fluid structure interaction occurs in the simulation behaviour of ocean-offshore platforms, aircraft flight characteristics, and reservoir dams. While there is a distinct existence and interaction between the solid and fluid in these issues, all these issues fall under the category of interaction between fluid and structure. It is also important to remember that the degree of magnitude of the solid and fluid interaction differs between the various problems. Although solid deformation is a part of many difficulties, there are many man-made problems in which the solid can be thought of as moving as a rigid body. One-directional interaction between the fluid and the solid is also conceivable in some cases. Fluid structure interaction problems, as well as multi-physics problems in general, are frequently too difficult to answer analytically, necessitating the use of experiments or numerical simulations. Work is still underway in the fields of computational fluid dynamics and computational structural dynamics, but the sophistication of these fields allows for numerical fluid structure interaction simulation.

2 Methods of FSI

In this paper, computational techniques that are currently available for computing fluid structure interaction problems based on conforming and non-conforming meshes, with a focus on recent developments in the field, are presented. One purpose is to categorise the methods chosen and to evaluate their accuracy and effectiveness.

2.1 Analysis of an Axial-Flow Pump

Irrigation and drainage systems have long used axial-flow pumps with a two-way path. Because of the geometry of the two-way inlet tube, the impeller easily vibrates due to chaotic turbulent flow. This vibration causes structural cracks and makes it difficult for the pump to operate safely. The results show that when the flow rate and head fall, the values of deformation and stress decrease rapidly, with the maximum total deformation and equivalent stress occurring around the impeller hub and rim, respectively. As illustrated in Fig. 1, the maximum stress occurred at the impeller tip. From blade leading edge to trailing edge, total deformations in the blade rim diminish, but corresponding stress in the blade hub increases at first and then declines, before

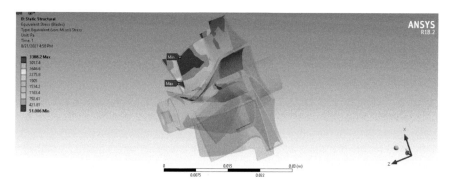

Fig. 1 Stress distribution on the impeller surface

rapidly increasing from blade outlet to inlet. At the hub, higher equivalent stress levels can be noticed, which quantitatively drop as the flow rate increases [1].

2.2 Performance Evaluation of a Membrane Blade

The goal of this project was to investigate the possibility of a membrane blade. In the sail wing principle, the wing surface, or blade, is made up of pre-tensioned membranes that meet at the trailing edge of the pre-tensioned edge cable. Because of the link between membrane deformation and applied aerodynamic load, a two-way coupled fluid structure interaction research is essential for determining the construction's aerodynamic efficiency. To handle the multi-physics challenge, CARAT++, an in-house finite-element-based structural solver, is integrated with Open FOAM. The membrane blade's lift coefficient, drag coefficient, and lift-to-drag ratio are compared to rigid equivalents. As a first step towards examining the definition for the rotating case, a single non-spinning NREL phase VI blade is examined. The membrane blade's lift curve is steeper than the rigid blade's. The lift and drag coefficients of the membrane blade, as well as the lift-to-drag ratio, are higher at higher angles of attack. As a preliminary step in determining the definition of the rotating event, a single non-spinning blade is examined [2].

To investigate the aerodynamic efficiency of a semi-flexible membrane blade, simulations of fluid structure interaction were performed under steady inflow conditions for a non-rotating blade. The length of the design blade under consideration is approximately 5 m. The leading edge is made up of a rigid mast, ribs surrounding the blade, tensioned edge cables at the trailing edge, and membranes that create the blade's upper and lower surfaces. In the absence of external force, the equilibrium shape of membrane structures is determined by the position of the supports and pretensions in the membranes, as well as the supporting edge cables. The equilibrium shape is determined via form-finding analysis. Internal forces and external loads, which are in turn governed by the actual shape, dictate the type of membrane structure for service conditions [3].

As a result, investigating this type of structures requires two-way coupled fluid structure interaction study. The vortex panel methodology and the numerical solution of the Navier–Stokes equations were both used to model the hydrodynamic problem. The finite-element method is used to analyse the structural problem nonlinearly. The purpose of this research was to discern between convergent FSI outcomes produced from two alternative techniques to tackling the fluid problem. This research is required before the adaptable blade's effective and exact multi-fidelity simulation idea for multiple design phases can be developed. The second purpose is to analyse and compare the membrane blade's aerodynamic efficiency to that of its rigid blade equivalent in terms of lift and drag coefficients, as well as lift-to-drag ratio. The blade configuration is taken from the NASA-Ames Process VI rotor as the rigid blade variant of the baseline. The stiff blade has a steeper lift curve slope and a higher lift-to-drag ratio than the membrane blade under investigation [3].

2.3   Analysis of Piezoelectric Flap

When microelectromechanical systems (MEMS) are used in a fluid medium, electric and fluid structure interaction (EFSI) is a difficult coupled multi-physics phenomenon. The electromechanical (electric-structure interaction) and fluid structure interaction coupling phenomena explored in this study are a combination of electromechanical (electric-structure interaction) and fluid structure interaction coupling. Both electromechanical coupling and fluid structural interaction can be modelled using a monolithic or partitioned iteration method [4].

The electromechanical coupling is simulated using the block Gauss–Seidel (BGS) iteration method in a partitioned iterative fashion with separate solvers for the electrical and mechanical equations, whereas the fluid structure interaction is simulated monolithically by solving the fluid and structure equations simultaneously using a projection method. To analyse the closely coupled EFSI in MEMS, the proposed strategy integrates these two approaches. The proposed method employs a flexible flap made up of a converging channel piezoelectric bimorph actuator. When a very low input bias voltage is applied to the actuator, the EFSI analysis results demonstrate a strong agreement with FSI outcomes [4].

2.4   Two-Way Fluid Structure Coupling

This study uses the commercial programme ANSYS-CFX to create a three-dimensional numerical fluid structure interaction model of a vibrating hydrofoil and studies hydrodynamic damping as a fluid contribution to total damping. In the FSI simulation, two independent solvers, one for the fluid (CFD) and one for the structure (FEM), were used, each running in sequential order with synchronisation points to transfer information at the fluid structure interface. This study uses the commercially available solvers ANSYS-CFX 13.0 and ANSYS Classic 13.0 as CFD and FEM solvers, respectively. Different meshes were generated for the fluid and solid fields in this study. Several factors, such as the flow velocity distribution around the blade, influence the identification of blade hydrodynamic damping, according to the findings. In order to compare the results of this research with those obtained through experimental observations and estimations under actual flowing conditions, the blade geometry was chosen to be identical to the hydrofoil blade utilised in ANDRITZ's experimental examination. The effects of using different flow velocities are also investigated [5].

3 Advantages of FSI

The interaction of a moveable or deformable structure with an internal or surrounding fluid flow is known as fluid structure interaction. Fluid structure interactions are an important consideration in the design of many engineering systems, and failure to consider the effects of oscillatory interactions, especially in structures built of materials, can be devastating.

3.1 Analysis on Wind Turbine Blade

The worldwide insights into breeze vitality show that the overall capacity to use wind management is evident. More than 51,477 MW of new wind generation capacity was installed in 2014; according to the Global Wind Energy Council (GWEC), by the end of 2014, the global total was over 369,553 MW [6].

As wind energy is one of the most promising forms, wind turbines have become the focus of energy-based research to harness this wind energy more effectively. The blade has been developed and simulated using ANSYS v16.0 and SST for a low twist angle to examine fluid flows around a typical horizontal axis wind turbine (HAWT) blade and 1-way fluid structure interaction (FSI) [7].

For free stream velocity of 12 m/s, 8 m/s, and 4 m/s, maximum 0.04 m, 0.05 m, and 0.07 m deflections were found, respectively. The stresses on the blade root for these velocities were 52 MPa, 73 MPa, and 99 MPa, respectively. It can be noted that the most deflection occurs at the 20–30% span distance from the tip and most stress occurs near the root only. Most of the deflection occurred in the tip and most of the stress occurred in the root section, whereas in other blades, generally stress occurred in the middle portion. The blade was designed for moderate power generation, and the power curve shows the linear increase in power for higher velocities. The power curve was plotted by the available torques at various velocities [7].

The aeroelastic response of a single wind turbine blade was studied using a computational model for fluid structure interaction analysis. The aerodynamic forces were calculated using the blade factor momentum (BEM) theory, which took into consideration the effects of wind shear and tower shadow. The blade's aeroelastic response was calculated by connecting these aerodynamic and structural models using a MATLAB-based coupled BEM–FEM technique. Aeroplane propeller blade studies can also benefit from this type of modelling [8].

3.2 Vibration of Gas Turbine Blade

It was investigated if casing wall pressure analysis might be used in the direct measurement of gas turbine rotor blade vibration amplitudes. The blade tip (arrival) time, which is the most popular non-contact measurement technology for gas turbine blade vibrations, is currently detected using many proximity probes positioned around the engine's periphery (BTT). Despite the system's growing capabilities, it still has major faults. These virtual pressure signals were used to perform computational simulations of casing wall pressures and reconstructions of rotor blade vibration amplitudes [9].

4 Application of FSI

Considering the fluid structure interaction challenges, the structural and fluid models must be merged or coupled. This paper explains the fundamental knowledge needed to construct and then evaluate a simple connection. The recommended method is to consider a dedicated solver for each of the two physical systems involved.

4.1 Pipeline System

According to a study of pipeline systems, pipelines carrying liquid have a significant impact in the construction area, especially in the fields of marine design, oil and gas, oil transportation, and city water supply. Pipelines that carry liquid are also thought to be the easiest way to solve liquid structure connection problems since they may show the pipeline's crucial dynamical conduct. Along these lines, the pipeline model is linked to the Euler–Bernoulli bar hypothesis, and the oil stream is modelled as a fitting stream model [10].

The fluid structure interaction (FSI) approach was used to determine the highest fatigue locations in blade structure for different manifolds. The maximum stress (Pa) on the turbocharger blade was the output parameter, while the input parameter was

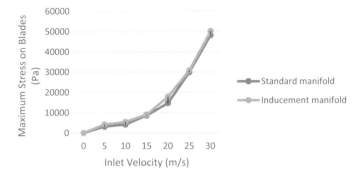

Fig. 2 FSI graphs for the standard manifold and the inducement manifold

the inlet velocity (m/s) of the manifold imported explicitly from the CFD computations. According to Fig. 2, the standard manifold and the inducement manifold had an average difference of 12% for inlet velocities ranging from 5 to 30 m/s. The inducement manifold contribution was slightly higher because of the design's influence on the increased inlet velocity.

4.2 FSI in Aerodynamics

A fluid structure interaction research was performed utilising three-dimensional Reynolds-averaged Navier–Stokes equations for a single-stage axial compressor with casing groove and tip injection. The blades were projected to distort based on the FSI study. The compressor's aerodynamic efficiency has been discovered to be influenced by blade deformation to some extent [11].

4.3 Aeroelastic Application

The interaction of a flexible structure with the surrounding fluid creates a range of phenomena that can be applied to wing stability analysis, turbomachinery design, bridge design, and arterial blood flow. Studying these phenomena involves both fluid and structure modelling. For aerodynamic and structural dynamic subsystems, many computational aeroelasticity approaches aim to combine independent computational approaches. Because of the interplay between the two simulation modules, this technique is known to be complicated. The goal is to find the best application-based fluid and structure models and connect them with an efficient interface. The availability of an effective moving grid technique to account for structural deformation is one of the most critical aspects of building a robust coupled aeroelastic model [8].

4.4 Aircraft Wing

Due to aerodynamic properties, the aircraft wing is also a complex structure to be tested and preserved for aeroelasticity. The problem of fluid structure interaction in the supercritical NASA SC (2)-0412 air foil was discussed in this paper. The main objective of this research was to find out the optimal output and deformation limit of the wing at various Mach numbers. This project is entirely carried out using CATIA and flow properties in the computational fluid dynamics (CFD) system by numerical methods of designing the wing. Finally, in ANSYS, the structural analysis is evaluated for deformation. The theoretical approach to the relationship between fluid and structure over an aircraft wing is complex [12].

5   Conclusions

The purpose of this work was to summarise the research on turbine fluid structure interaction. The need of obtaining reliable data on critical parameters, flow-induced excitation, added mass effect, hydrodynamic damping, blade flutter, and fatigue loading is all discussed. Fluid structure interaction is a subdiscipline focused on mathematical explanations and numerical approximations of the interactions between fluids and structures in computational mechanics, engineering, and science.

Acknowledgements The authors thank the Ministry of Higher Education (MOHE), Malaysia, for the financial support provided via Fundamental Research Grant Scheme [ref no FRGS/1/2019/TK03/UNIKL/01/2]. The authors also thank Universiti Kuala Lumpur, Malaysia, for the support provided via UniKL-STRG Grant [ref no str17042], and UniKL Malaysian Spanish Institute for providing necessary facilities and resources to complete this study for publication.

References

1. F.M. Ji Pei, Fluid–structure coupling analysis of deformation and stress in impeller of an axial-flow pump with two-way passage. Adv. Mech. Eng. **8**(4), 1–11 (2016)
2. M. Saeed, Fluid-Structure interaction analysis and performance evaluation of a membrane blade. Sci. Mak. Torque Win **753** (2016). https://doi.org/10.1088/1742-6596/753/10/102009
3. K.-U.B. Mehran Saeedi, Multi-fidelity fluid–structure interaction analysis of a membrane blade concept in non-rotating, uniform flow condition. Wind Energy Sci. 255–269 (2016)
4. D.S. Prakasha Chigahalli Ramegowda Fluid-structure and electric interaction analysis of piezo-electric flap, in *A Channel Using A Strongly Coupled FEM Scheme. Department of Mechanical Information Science and Technology* (2018)
5. T. Liaghat, Two-way fluid-structure coupling in vibration and damping analysis of an oscillating hydrofoil (2014)
6. P. Ravi Kumar, K. Awadheh, Fluid-structure interaction analysis on horizontal wind turbine blade. IJMPERD **12**(9), 1467 (2020)

7. M.M. Sumaiya Tasnim, A Numerical study on the fluid structure interaction of a wind turbine blade. Int. J. Mech. Mater. Eng. (2017). Paper ID : FM-244, RUET, Rajshahi, Bangladesh
8. A. Mauwafak, M.I.T. Tawfik, Aeroelastic behavior of a wind turbine blade by a fluid -structure interaction analysis. Al-Khwarizmi Eng. J. **1**, 15–25 (2013)
9. Forbes, G.L., Alshroof, O.N., Randall, R.B.: Fluid-structure interaction study of gas turbine blade vibrations. 6th Australasian Congress on Applied Mechanics, ACAM 6, 12–15 December 2010, Perth, Australia (2010)
10. K.S. Fong, M.Y. Airil Yasreen, Fluid-structure interaction (FSI) of damped oil conveying pipeline system by finite element method. MATEC Web Conf **111**, 01005 (2017)
11. S.B. Ma, K.Y. Kim, M.W. Hoe, J. Choi, Aerodynamic investigation of a single-stage axial compressor with a casing groove and tip injection using fluid-structure interaction analysis. J. Proc. ASME Turbo. Expo. (2015)
12. P.S. Jain, K.P. Gowda, Fluid-structure interaction over an aircraft wing structure. IRJET **8**(5), 2516–2527 (2021)

Printed in the United States
by Baker & Taylor Publisher Services